# Contents

### Chapter 1: Causes of Extinction

| | |
|---|---|
| What is mass extinction and are we facing a sixth one? | 1 |
| The five biggest threats to our natural world...and how we can stop them | 3 |
| 5 threats to UK wildlife | 8 |
| The world is in trouble: one million animals and plants face extinction | 10 |
| Biodiversity: we can map the biggest threats to endangered species in your local area | 12 |
| Wildlife trafficking driving 'severe declines' in traded species, finds study | 14 |
| Five reasons people buy illegal wildlife products – and how to stop them | 16 |

### Chapter 2: Species at Risk

| | |
|---|---|
| One quarter of native mammals now at risk of extinction in Britain | 18 |
| Only two northern white rhinos remain, and they're both female – here's how we could make more | 20 |
| 'Really sad moment': bogong moth among 124 Australian additions to endangered species list | 21 |
| New UK Red List for birds: more than one in four species in serious trouble | 22 |
| Illegal ivory trade shrinks while pangolin trafficking booms, UN says | 24 |

### Chapter 3: Hope for the Future?

| | |
|---|---|
| How hybrids could help save endangered species | 25 |
| How lockdown has helped the world's endangered species bounce back | 27 |
| Sharks will soon have their own 'superhighway' in the Pacific Ocean | 29 |
| Beavers born in Essex 'first time since the Middle Ages' | 30 |
| Historic reintroduction reverses extinction of England's rarest frog | 31 |
| Dogs are saving endangered wildlife in Africa | 32 |
| 200th osprey chick for pioneering Rutland Osprey Chick | 34 |
| The first endangered American animal has been cloned | 36 |
| UK extinct species rediscovered in the Outer Hebrides | 37 |
| At least 28 extinctions prevented by conservation action | 38 |

| | |
|---|---|
| Key Facts | 40 |
| Glossary | 41 |
| Activities | 42 |
| Index | 43 |
| Acknowledgements | 44 |

# Introduction

Endangered Species is Volume 402 in the **issues** series. The aim of the series is to offer current, diverse information about important issues in our world, from a UK perspective.

## ABOUT ENDANGERED SPECIES

Human activity is destroying nature at an unprecedented rate. Around one million animals and plants currently face the threat of extinction. This book examines the causes and consequences of the erosion of our natural world and the many species at risk today. It also looks at what is being done to protect endangered species as well as sharing some positive stories of hope.

## OUR SOURCES

Titles in the **issues** series are designed to function as educational resource books, providing a balanced overview of a specific subject.

The information in our books is comprised of facts, articles and opinions from many different sources, including:

- Newspaper reports and opinion pieces
- Website factsheets
- Magazine and journal articles
- Statistics and surveys
- Government reports
- Literature from special interest groups.

## A NOTE ON CRITICAL EVALUATION

Because the information reprinted here is from a number of different sources, readers should bear in mind the origin of the text and whether the source is likely to have a particular bias when presenting information (or when conducting their research). It is hoped that, as you read about the many aspects of the issues explored in this book, you will critically evaluate the information presented.

It is important that you decide whether you are being presented with facts or opinions. Does the writer give a biased or unbiased report? If an opinion is being expressed, do you agree with the writer? Is there potential bias to the 'facts' or statistics behind an article?

## ASSIGNMENTS

In the back of this book, you will find a selection of assignments designed to help you engage with the articles you have been reading and to explore your own opinions. Some tasks will take longer than others and there is a mixture of design, writing and research-based activities that you can complete alone or in a group.

## FURTHER RESEARCH

At the end of each article we have listed its source and a website that you can visit if you would like to conduct your own research. Please remember to critically evaluate any sources that you consult and consider whether the information you are viewing is accurate and unbiased.

# Useful Websites

www.arc-trust.org

www.buglife.org.uk

www.independent.co.uk

www.mammal.org.uk

www.ncl.ac.uk

www.news-decoder.com

www.nhm.ac.uk

www.positive.news

www.rarebirdalert.co.uk

www.singularityhub.com

www.telegraph.co.uk

www.theconversation.com

www.theguardian.com

www.wildlifetrusts.org

www.wwf.org.uk

# Chapter 1: Casues of Extinction

## What is mass extinction and are we facing a sixth one?

Human activity is killing nature at an unprecedented rate. We are now experiencing the consequences in the form of a possible sixth mass extinction.

By Tammana Begum

### The definition of a mass extinction

Extinction is a part of life, and animals and plants disappear all the time. About 98% of all the organisms that have ever existed on our planet are now extinct.

When a species goes extinct, its role in the ecosystem is usually filled by new species, or other existing ones. Earth's 'normal' extinction rate is often thought to be somewhere between 0.1 and 1 species per 10,000 species per 100 years. This is known as the background rate of extinction.

A mass extinction event is when species vanish much faster than they are replaced. This is usually defined as about 75% of the world's species being lost in a 'short' amount of geological time - less than 2.8 million years.

Katie Collins, Curator of Benthic molluscs at the [Natural History] Museum says, 'It's difficult to identify when a mass extinction may have started and ended. However, there are five big events that we know of, where extinction was much higher than normal background rate, and these are often used to decide whether we are going through a sixth one now.'

### How many mass extinctions have there been?

Five great mass extinctions have changed the face of life on Earth. We know what caused some of them, but others remain a mystery.

1. **The Ordovician-Silurian mass extinction** occurred 443 million years ago and wiped out approximately 85% of all species. Scientists think it was caused by temperatures plummeting and huge glaciers forming, which caused sea levels to drop dramatically. This was followed by a period of rapid warming. Many small marine creatures died out.

2. **The Devonian mass extinction** event took place 374 million years ago and killed about three-quarters of the world's species, most of which were marine invertebrates that lived at the bottom of the sea. This was a period of many environmental changes, including global warming and cooling, a rise and fall of sea levels and a reduction in oxygen and carbon dioxide in the atmosphere. We don't know exactly what triggered the extinction event.

3. **The Permian mass extinction**, which happened 250 million years ago, was the largest and most devastating event of the five. Also known as the Great Dying, it eradicated more than 95% of all species, including most of the vertebrates which had begun to evolve by this time. Some scientists think Earth was hit by a large asteroid which filled the air with dust particles that blocked out the Sun and caused acid rain. Others think there was a large volcanic explosion which increased carbon dioxide and made the oceans toxic.

4. **The Triassic mass extinction** event took place 200 million years ago, eliminating about 80% of Earth's species,

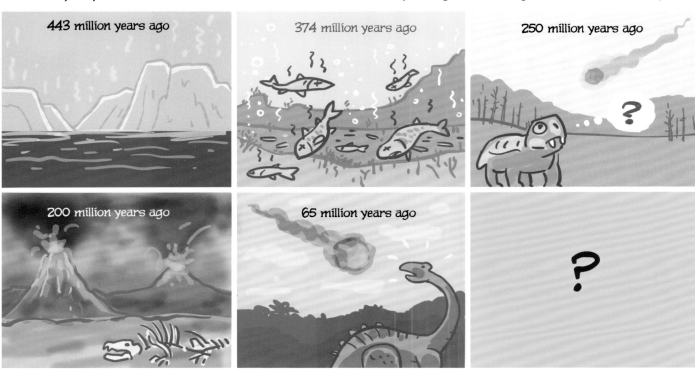

*issues:* Endangered Species

including many types of dinosaurs. This was probably caused by colossal geological activity that increased carbon dioxide levels and global temperatures, as well as ocean acidification.

*5. The Cretaceous mass extinction* event occurred 65 million years ago, killing 78% of all species, including the remaining non-avian dinosaurs. This was most likely caused by an asteroid hitting the Earth in what is now Mexico, potentially compounded by ongoing flood volcanism in what is now India.

## What causes mass extinctions?

Past mass extinctions were caused by extreme temperature changes, rising or falling sea levels and catastrophic, one-off events like a huge volcano erupting or an asteroid hitting Earth.

We know about them because we can see how life has changed in the fossil record. For instance, a large part of Katie's work includes exploring extinction through fossils such as bivalves.

Katie says, 'Bivalves have been around for 500 million years, making them one of the oldest groups of fossils we can study and still see how they live and survive today. We get some really good continuous data from them all around the world.'

While fossils can tell us a lot about how life used to be on Earth, there are still many questions that remain unanswered.

'The Cretaceous-Paleogene extinction is the youngest mass extinction event, and probably the most studied,' Katie adds. 'We should understand the Cretaceous event pretty well, but many aspects of it, including the lead-in, the cause and the recovery, are all still areas of active research.'

## Are we living through the sixth mass extinction?

We are experiencing drastic changes to our planet, including extreme weather such as flooding, drought and wildfires.

Research, including some led by the Museum, shows humans are the cause of these changes. Since the Industrial Revolution, we have been putting pressure on nature by using its resources without supporting recovery.

For example, land use change is continuing to destroy swathes of natural landscapes. Humans have already transformed over 70% of land surfaces and are using about three-quarters of freshwater resources.

Agriculture is also a leading cause of soil degradation, deforestation, pollution and biodiversity loss. It is diminishing wild spaces and driving out countless species from their natural habitats, forcing them to clash with humans for resources or leaving them vulnerable.

Katie adds, 'Many large animals are culled because they are seen as a risk to humans. People will hunt predatory birds disproportionally as they consider them a threat to farming, although they mostly eat rabbits.

'A lot of wolves have been removed in North America because they're seen as predators of livestock and that's caused a trophic ecological cascade.'

Cane toads are a notorious invasive species in Australia. They breed year-round and a female can lay up to 35,000 eggs at one time, and can thrive in a variety of environments.

Invasive species, many of which are introduced by humans, are also threatening ecosystems all over the world. Introduced species compete with local species for resources and often diminish the quality of biodiversity in the area, sometimes causing extinction. These are just some of the devastating changes caused by humans.

All life on Earth is finely interwoven. This delicate balance has been established over millions of years. As one species becomes extinct, many other species are affected, putting a number of ecosystems in danger of collapsing.

Naturally, extinction occurs over hundreds and thousands of years which allows nature to slowly replace what has been lost. But humans have sped up this process to a dangerous rate.

Katie says, 'The current rate of extinction is between 100 and 1,000 times higher than the pre-human background rate of extinction, which is jaw-dropping. We are definitely going through a sixth mass extinction.'

Never before has a single species been responsible for such destruction on Earth.

## Could we stop a sixth mass extinction?

Mass extinctions are a large and complex issue. They can be slow burners, taking millions of years to unfold.

Right now, it seems likely we are experiencing a sixth, and it is undoubtably the result of human actions, including human-induced climate change.

'The floods and wildfires we're hearing about in the news now will become regular occurrences in 50 years' time,' says Katie. 'They will test the resilience of our buildings, infrastructures, transatlantic cables, satellites and more.

'These natural disasters are going to exacerbate existing inequalities, but it doesn't have to be that way. Research shows that if we change how we use natural resources now, the future could be a positive one for the next generation.

Katie says, 'If we can work on reducing the negative impact we've had on the climate, then other things will also improve, such as the number of species that are currently threatened by habitat loss.

'We need to work on how we access and use natural resources, including land management. Habitat loss is a huge problem and land use is tied in with that.'

Many believe the changes we need to see now can be achieved fastest by prioritising the protection and preservation of nature over the interests of financial systems.

Katie says, 'I know there is a lot of emphasis on individual action but most of the climate-altering pollution and fossil fuel burning is the responsibility of a small number of parties.

'It would be much more effective for individuals to put pressure on policymakers and businesses to reduce emissions and target companies that are major emitters.'

The future of our world hangs on our making what is perhaps the biggest international effort in history to reduce human impacts. We all have an active role to play, which requires deep transformation of our values, attitudes and behaviours.

*19 May 2021*

The above information is reprinted with kind permission from *NHM*
© The Trustees of The Natural History Museum, London 2021

www.nhm.ac.uk

# The five biggest threats to our natural world ... and how we can stop them

From destructive land use to invasive species, scientists have identified the main drivers of biodiversity loss – so that countries can collectively act to tackle them.

By Patrick Greenfield and Phoebe Weston

The world's wildlife populations have plummeted by more than two-thirds since 1970 – and there are no signs that this downward trend is slowing. The first phase of Cop15 talks in Kunming this week will lay the groundwork for governments to draw up a global agreement next year to halt the loss of nature. If they are to succeed, they will need to tackle what the IPBES (Intergovernmental Science-Policy Platform on Biodiversity and Ecosystem Services) has identified as the five key drivers of biodiversity loss: changes in land and sea use; direct exploitation of natural resources; climate change; pollution; and invasion of alien species.

## Changes in land and sea use

'It's hidden destruction. We're still losing grasslands in the US at a rate of half a million acres a year or more.'

Tyler Lark, from the University of Wisconsin-Madison, knows what he is talking about. Lark and a team of researchers used satellite data to map the expansion and abandonment of land across the US and discovered that 4m hectares (10m acres) had been destroyed between 2008 and 2016.

Large swathes of the United States' great prairies continue to be converted into cropland, according to the research, to make way for soya bean, corn and wheat farming.

Changes in land and sea use has been identified as the main driver of 'unprecedented' biodiversity and ecosystem change over the past 50 years. Three-quarters of the land-based environment and about 66% of the marine environment have been significantly altered by human actions.

North America's grasslands – often referred to as prairies – are a case in point. In the US, about half have been converted since European settlement, and the most fertile land is already being used for agriculture. Areas converted more recently are sub-prime agricultural land, with 70% of yields lower than the national average, which means a lot of biodiversity is being lost for diminishing returns.

'Our findings demonstrate a pervasive pattern of encroachment into areas that are increasingly marginal for production but highly significant for wildlife,' Lark and his team wrote in the paper, published in Nature Communications.

Boggier areas of land, or those with uneven terrain, were traditionally left as grassland, but in the past few decades, this marginal land has also been converted. In the US, 88% of cropland expansion takes place on grassland, and much of this is happening in the Great Plains – known as America's breadbasket – which used to be the most extensive grassland in the world.

## What are the five biggest threats to biodiversity?

Hotspots for this expansion have included wildlife-rich grasslands in the 'prairie pothole' region which stretches between Iowa, Dakota, Montana and southern Canada and is home to more than 50% of North American migratory waterfowl, as well as 96 species of songbird. This cropland expansion has wiped out about 138,000 nesting habitats for waterfowl, researchers estimate.

These grasslands are also a rich habitat for the monarch butterfly – a flagship species for pollinator conservation and a key indicator of overall insect biodiversity. More than 200m milkweed plants, the caterpillar's only food source, were probably destroyed by cropland expansion, making it one of the leading causes for the monarch's national decline.

The extent of conversion of grassland in the US makes it a larger emission source than the destruction of the Brazilian Cerrado, according to research from 2019. About 90% of emissions from grassland conversion comes from carbon lost in the soil, which is released when the grassland is ploughed up.

'The rate of clearing that we're seeing on these grasslands is on par with things like tropical deforestation, but it often receives far less attention,' says Lark.

Food crop production globally has increased by about 300% since 1970, despite the negative environmental impacts.

Reducing food waste and eating less meat would help cut the amount of land needed for farming, while researchers say improved management of existing croplands and utilising what is already farmed as best as possible would reduce further expansion.

Lark concludes: 'I think there's a huge opportunity to re-envision our landscapes so that they're not only providing incredible food production but also mitigating climate change and helping reduce the impacts of the biodiversity crisis by increasing habitats on agricultural land.'

## Direct exploitation of natural resources

*Groundwater extraction: 'People don't see it'*

From hunting, fishing and logging to the extraction of oil, gas, coal and water, humanity's insatiable appetite for the planet's resources has devastated large parts of the natural world.

While the impacts of many of these actions can often be seen, unsustainable groundwater extraction could be driving a hidden crisis below our feet, experts have warned, wiping out freshwater biodiversity, threatening global food security and causing rivers to run dry.

Farmers and mining companies are pumping vast underground water stores at an unsustainable rate, according to ecologists and hydrologists. About half the world's population relies on groundwater for drinking water and it helps sustain 40% of irrigation systems for crops.

The consequences for freshwater ecosystems – among the most degraded on the planet – are under-researched as studies have focused on the depletion of groundwater for agriculture.

But a growing body of research indicates that pumping the world's most extracted resource – water – is causing significant damage to the planet's ecosystems. A 2017 study of the Ogallala aquifer – an enormous water source underneath eight states in the US Great Plains – found that more than half a century of pumping has caused streams to run dry and a collapse in large fish populations. In 2019, another study estimated that by 2050 between 42% and 79% of watersheds that pump groundwater globally could pass ecological tipping points, without better management.

'The difficulty with groundwater is that people don't see it and they don't understand the fragility of it,' says James Dalton, director of the global water programme at the International Union for Conservation of Nature (IUCN).

'Groundwater can be the largest – and sometimes the sole – source in certain types of terrestrial habitats.

'Uganda is luxuriantly green, even during the dry season, but that's because a lot of it is irrigated with shallow groundwater for agriculture and the ecosystems are reliant on tapping into it.'

According to UPGro (Unlocking the Potential of Groundwater for the Poor), a research programme looking into the management of groundwater in sub-Saharan Africa, 73 of the 98 operational water supply systems in Uganda are dependent on water from below ground. The country shares two transboundary aquifers: the Nile and Lake Victoria basins. At least 592 aquifers are shared across borders around the world.

'Some of the groundwater reserves are huge, so there is time to fix this,' says Dalton. 'It's just there's no attention to it.'

Inge de Graaf, a hydrologist at Wageningen University, who led the 2019 study into watershed levels, found between 15% to 21% had already passed ecological tipping points, adding that once the effects had become clear for rivers, it was often too late.

'Groundwater is slow because it has to flow through rocks. If you extract water today, it will impact the stream flow maybe in the next five years, in the next 10 years, or in the next decades,' she says. 'I think the results of this research and related studies are pretty scary.'

In April, the largest ever assessment of global groundwater wells by researchers from University of California, Santa Barbara, found that up to one in five were at risk of running dry. Scott Jasechko, a hydrologist and lead author on the paper, says that the study focuses on the consequences for humans and more research is needed on biodiversity.

'Millions of wells around the world could run dry with even modest declines in groundwater levels. And that, of course, has cascading implications for livelihoods and access to reliable and convenient water for individuals and ecosystems,' he says.

## The climate crisis

### Climate and biodiversity: 'Solve both or solve neither'

In 2019, the European heatwave brought 43C heat to Montpellier in France. Great tit chicks in 30 nest boxes starved to death, probably because it was too hot for their parents to catch the food they needed, according to one researcher. Two years later, and 2021's heatwave appears to have set a European record, pushing temperatures to 48.8C in Sicily in August. Meanwhile, wildfires and heatwaves are stripping the planet of life.

Until now, the destruction of habitats and extraction of resources has had a more significant impact on biodiversity than the climate crisis. This is likely to change over the coming decades as the climate crisis dismantles ecosystems in unpredictable and dramatic ways, according to a review paper published by the Royal Society.

'There are many aspects of ecosystem science where we will not know enough in sufficient time,' the paper says.

'Ecosystems are changing so rapidly in response to global change drivers that our research and modelling frameworks are overtaken by empirical, system- altering changes.'

The calls for biodiversity and the climate crisis to be tackled in tandem are growing. 'It is clear that we cannot solve [the global biodiversity and climate crises] in isolation – we either solve both or we solve neither,' says Sveinung Rotevatn, Norway's climate and environment minister, with the launch in June of a report produced by the world's leading biodiversity and climate experts. Zoological Society of London senior research fellow Dr Nathalie Pettorelli, who led a study on the subject published in the Journal of Applied Ecology in September, says: 'The level of interconnectedness between the climate change and biodiversity crises is high and should not be underestimated. This is not just about climate change impacting biodiversity; it is also about the loss of biodiversity deepening the climate crisis.'

Writer Zadie Smith describes every country's changes as a 'local sadness'. Insects no longer fly into the house when the lights are on in the evening, the snowdrops are coming out earlier and some migratory species, such as swallows, are starting to try to stay in the UK for winter. All these individual elements are entwined in a much bigger story of decline.

Our biosphere – the thin film of life on the surface of our planet – is being destabilised by temperature change. On land, rains are altering, extreme weather events are more common, and ecosystems more flammable. Associated changes, including flooding, sea level rise, droughts and storms, are having hugely damaging impacts on biodiversity and its ability to support us.

In the ocean, heatwaves and acidification are stressing organisms and ecosystems already under pressure due to other human activities, such as overfishing and habitat fragmentation.

The latest Intergovernmental Panel on Climate Change (IPCC) landmark report showed that extreme heatwaves that would usually happen every 50 years are already happening every decade. If warming is kept to 1.5C these will happen approximately every five years.

The distributions of almost half (47%) of land-based flightless mammals and almost a quarter of threatened birds, may already have been negatively affected by the climate crisis, the IPBES warns. Five per cent of species are at risk of extinction from 2C warming, climbing to 16% with a 4.3C rise.

Connected, diverse and extensive ecosystems can help stabilise the climate and will have a better chance of thriving in a world permanently altered by rising emissions, say experts. And, as the Royal Society paper says: 'Rather than being framed as a victim of climate change, biodiversity can be seen as a key ally in dealing with climate change.'

## Pollution

### The hidden threat of nitrogen: 'Slowly eating away at biodiversity'

On the west coast of Scotland, fragments of an ancient rainforest that once stretched along the Atlantic coast of Britain cling on. Its rare mosses, lichens and fungi are perfectly suited to the mild temperatures and steady supply of rainfall, covering the crags, gorges and bark of native woodland. But nitrogen pollution, an invisible menace, threatens the survival of the remaining 30,000 hectares (74,000 acres) of Scottish rainforest, along with invasive rhododendron, conifer plantations and deer.

While marine plastic pollution in particular has increased tenfold since 1980 – affecting 44% of seabirds – air, water and soil pollution are all on the rise in some areas. This has led to pollution being singled out as the fourth biggest driver of biodiversity loss.

In Scotland, nitrogen compounds from intensive farming and fossil fuel combustion are dumped on the Scottish rainforest from the sky, killing off the lichen and bryophytes that absorb water from the air and are highly sensitive to atmospheric conditions.

'The temperate rainforest is far from the sources of pollution, yet because it's so rainy, we're getting a kind of acid rain effect,' says Jenny Hawley, policy manager at Plantlife, which has called nitrogen pollution in the air 'the elephant in the room' of nature conservation. 'The nitrogen-rich rain that's coming down and depositing nitrogen into those habitats is making it impossible for the lichen, fungi, mosses and wildflowers to survive.'

Environmental destruction caused by nitrogen pollution is not limited to the Scottish rainforest. Algal blooms around the world are often caused by runoff from farming, resulting in vast dead zones in oceans and lakes that kill scores of fish and devastate ecosystems. Nitrogen-rich rainwater degrades the ability of peatlands to sequester carbon, the protection of which is a stated climate goal of several governments. Wildflowers adapted to low-nitrogen soils are squeezed out by aggressive nettles and cow parsley, making them less diverse.

About 80% of nitrogen used by humans – through food production, transport, energy and industrial and wastewater processes – is wasted and enters the environment as pollution.

### 'In terms of a nitrogen footprint, the most intensive thing that you can eat is meat'

'Nitrogen pollution might not result in huge floods and apocalyptic droughts but we are slowly eating away at biodiversity as we put more and more nitrogen in ecosystems,' says Carly Stevens, a plant ecologist at Lancaster University. 'Across the UK, we have shown that habitats that have lots of nitrogen have fewer species in them. We have shown it across Europe. We have shown it across the US. Now we're showing it in China. We're creating more and more damage all the time.'

To decrease the amount of nitrogen pollution causing biodiversity loss, governments will commit to halving nutrient runoff by 2030 as part of an agreement for nature currently being negotiated in Kunming. Halting the waste of vast amounts of nitrogen fertiliser in agriculture is a key part of meeting the target, says Kevin Hicks, a senior research fellow at the Stockholm Environment Institute centre at York.

'One of the biggest problems is the flow of nitrogen from farming into watercourses,' Hicks says. 'In terms of a nitrogen footprint, the most intensive thing that you can eat is meat. The more meat you eat, the more nitrogen you're putting into the environment.'

Mark Sutton, a professor at the UK Centre for Ecology & Hydrology, says reducing nitrogen pollution also makes economic sense.

'Nitrogen in the atmosphere is 78% of every breath we take. It does nothing, it's very stable and makes the sky blue. Then there are all these other nitrogen compounds: ammonia, nitrates, nitrous oxide. They create air and water pollution,' he says. He argues that if you price every kilo of nitrogen at $1 (an estimated fertiliser price), and multiply it by the amount of nitrogen pollution lost in the world – 200bn tonnes – it amounts to $200bn (£147bn) every year.

'The goal to cut nitrogen waste in half would save you $100bn,' he says. 'I think $100bn a year is a worthwhile saving.'

## Invasive species

### The problem for islands: 'We have to be very careful'

On Gough Island in the southern Atlantic Ocean, scores of seabird chicks are eaten by mice every year. The rodents were accidentally introduced by sailors in the 19th century and their population has surged, putting the Tristan albatross – one of the largest of its species – at risk of extinction along with dozens of rare seabirds. Although Tristan albatross chicks are 300 times the size of mice, two-thirds did not fledge in 2020 largely because of the injuries they sustained from the rodents, according to the RSPB.

The situation on the remote island, 2,600km from South Africa, is a grisly warning of the consequences of the human-driven impacts of invasive species on biodiversity. An RSPB-led operation to eradicate mice from the British overseas territory has been completed, using poison to help save the critically endangered albatross and other bird species from injuries they sustain from the rodents. It will be two years before researchers can confirm whether or not the plan has worked. But some conservationists want to explore another controversial option whose application is most advanced in the eradication of malaria: gene drives.

Instead of large-scale trapping or poisoning operations, which have limited effectiveness and can harm other species, gene drives involve introducing genetic code into an invasive population that would make them infertile or all one gender over successive generations. The method has so far been used only in a laboratory setting but at September's IUCN congress in Marseille, members backed a motion to develop a policy on researching its application and other uses of synthetic biology for conservation.

'If a gene drive were proven to be effective and there were safety mechanisms to limit its deployment, you would introduce multiple individuals on an island whose genes would be inherited by other individuals in the population,' says David Will, an innovation programme manager with Island Conservation, a non-profit dedicated to preventing extinctions by removing invasive species from islands. 'Eventually, you would have either an entirely all male or entirely all female population and they would no longer be able to reproduce.'

Nearly one-fifth of the Earth's surface is at risk of plant and animal invasions and although the problem is worldwide, such as feral pigs wreaking havoc in the southern United States and lionfish in the Mediterranean, islands are often worst affected. The global scale of the issue will be revealed in a UN scientific assessment in 2023.

'We have to be very careful,' says Austin Burt, a professor of evolutionary genetics at Imperial College London, who researches how gene drives can be used to eradicate malaria in mosquito populations. 'If you're going after mice, for example, and you're targeting mice on an island, you'd need to make sure that none of those modified mice got off the island to cause harm to the mainland population.'

In July, scientists announced they had successfully wiped out a population of malaria-transmitting mosquitoes using a gene drive in a laboratory setting, raising the prospect of self-destructing mosquitoes being released into the wild in the next decade.

Kent Redford, chair of the IUCN Task Force on Synthetic Biology who led an assessment of the use of synthetic biology in conservation, said there are clear risks and opportunities in the field but further research is necessary.

'None of these genetic tools are ever going to be a panacea. Ever. Nor do I think they will ever replace the existing tools,' Redford says, adding: 'There is a hope – and I stress hope – that engineered gene drives have the potential to effectively decrease the population sizes of alien invasive species with very limited knock-on effects on other species.'

*14 October 2021*

The above information is reprinted with kind permission from *The Guardian*.
© 2022 Guardian News and Media Limited

www.theguardian.com

# 5 threats to UK wildlife

We need to act quickly to restore nature to its glorious best.

It's the perfect time to get outdoors and enjoy our Great British countryside – including the amazing wildlife that call it home.

But we're losing our incredible wildlife at an alarming rate. Between 1970 and 2013, 56% of species in the UK declined. We need to act quickly to restore nature to its glorious best.

Here we look at 5 threats to UK wildlife.

## 1. Agricultural intensification

### The problem

The biggest impact on UK wildlife has been the intensification of agriculture.

Soil health has deteriorated. Numbers of farmland bird species such as the grey partridge, tree sparrow, skylark, linnet and yellowhammer have dropped. Precious UK habitats have been eroded.

Intensive farming has resulted in the loss of flower meadows, hedgerows and trees – all of which are vital habitats for pollinating insects such as bees, with knock-on effects for species further up the food chain.

### What we can do about it

We need to make sure farmers are supported to do things that are good for nature but don't bring in direct income.

As we leave the EU, we have a golden opportunity to change the way we manage our land in bold ways that could help us restore nature.

## 2. Plastics

### The problem

It's been hard to miss some of the shocking facts about our plastic waste recently. Did you know for instance, that by 2050 there could be more plastic in the ocean than fish? Or that we flush away 3.4 billion wet wipes each year?!

Plastics are choking our rivers and seas and killing our precious wildlife – including here in the UK. Seabirds are found with their stomachs full of plastic items, while microplastics are consumed by animals such as plankton, passing the problem back up the food chain.

### What we can do about it

Individuals, businesses and governments all have an important role to play in stemming the plastic tide.

## 3. Climate change

### The problem

You're probably aware of the threat that climate change poses to iconic species such as polar bears or snow leopards. But it will increasingly have an impact on wildlife right here in the UK too.

issues: Endangered Species — Chapter 1: Causes of Extinction

Changing weather patterns – for example longer periods of drought, warmer summers, or more flooding – will all have a knock-on effect on our wildlife. Many of our most iconic mammals, such as badgers, moles and hedgehogs, eat invertebrates that favour wetter weather and could be threatened by climate change.

Many migratory birds could also be affected, as their passage is closely synchronised with food availability. Shifts in seasons could therefore have a serious impact on their survival chances.

*What we can do about it*

We have the knowledge and the technology to reduce our impact on the climate and ease the pressures on the world's most vulnerable places, people and wildlife. We just need to make it happen.

## 4. River Damage

*The problem*

Our rivers not only help make the British landscape so picturesque and vibrant, they're also a vital source of fresh water for people, industry, farming and wildlife.

But our rivers have suffered from 'over-abstraction' – taking too much water out – as well as pollution from fertilisers and pesticides. Less than a fifth of England's rivers are healthy, which poses a big threat to amazing UK wildlife such as water voles and kingfishers.

*What we can do about it*

Improve farming practices and reduce agricultural pollution.

## 5. Pesticides

*The problem*

Another result of the intensification of agriculture has been an increase in the use of pesticides. Specific types of pesticides – neonicotinoids – were banned by the EU due to concern about the large scale harm caused to bees.

But they're not the only pesticides that pose a threat to UK wildlife. In fact, many slug pellets used in gardens around Britain contain an active ingredient called 'metaldehyde' (a chemical also used widely by farmers) which can cause huge damage if it's not applied carefully. It can enter our waterways and into our drinking water.

The pellets can also be washed into drains or ditches, find their way into our rivers, and harm animals much larger than slugs. Like microplastics, this toxic chemical can be passed up the food chain, harming predators like hedgehogs or birds.

*The solution*

Try to avoid metaldehyde, using natural alternatives such as copper strips to deter slugs – or look to create an environment that encourages predators like slow-worms or hedgehogs.

*2021*

The above information is reprinted with kind permission from WWF
© 2022 WWF-UK

www.wwf.org.uk

# The world is in trouble: one million animals and plants face extinction

A landmark report has confirmed that humanity is destroying its own life support system as the natural world faces unprecedented declines.

An international team of scientists, backed by the UN, has reported that communities around the world are likely to face dire consequences as ecosystems decline faster and faster.

By Katie Pavid

Human impacts on the natural environment are now so great that we are eroding our own economies and food security, according to the world's leading climate scientists.

Every ecosystem around the world is affected by extinction, from coral reefs to tropical jungles, and the problem is accelerating with each passing day.

It is estimated that around one million animals and plants are threatened with extinction - more than ever before in human history. More than 40% of amphibian species, about 33% of reef-forming corals and more than a third of all marine mammals are threatened.

And it is humanity that is to blame, as about 75% of environments on land have been significantly altered by human actions, plus roughly 66% of the marine environment.

Coral reefs are home to almost a quarter of all marine species. Roughly 33% of reef-forming corals are now threatened with extinction.

Compiled by 145 expert authors from 50 countries, with inputs from another 310 contributing authors, the report was released by the Intergovernmental Science-Policy Platform on Biodiversity and Ecosystem Services (IPBES).

It is the most comprehensive planetary health-check of its kind, having examined changes to the natural world over the past five decades.

Experts studied the relationship between economic development and human impact on nature. They also offered a range of possible scenarios for the coming decades and found that action is needed urgently if we are to protect both people and the planet from catastrophic damage.

Sir Robert Watson is one of the most influential environmental scientists in the world and chairs the IPBES, an independent intergovernmental body formed of more than 130 member governments.

He says, 'The overwhelming evidence of the IPBES Global Assessment, from a wide range of different fields of knowledge, presents an ominous picture.

'The health of ecosystems on which we and all other species depend is deteriorating more rapidly than ever. We are eroding the very foundations of our economies, livelihoods, food security, health and quality of life worldwide.

'The report also tells us that it is not too late to make a difference, but only if we start now at every level from local to global. Through "transformative change", nature can still be conserved, restored and used sustainably - this is also key to meeting most other global goals. By transformative change, we mean a fundamental, system-wide reorganization across technological, economic and social factors, including paradigms, goals and values.'

## A human problem

The IPBES Global Assessment Report on Biodiversity and Ecosystem Services is the most comprehensive ever completed. It is the first intergovernmental report of its kind and builds on the Millennium Ecosystem Assessment of 2005.

Scientists drew together and reviewed 15,000 scientific and government sources, as well as studying issues that are directly affecting indigenous peoples and local communities.

Based on the evidence available, the authors ranked the five biggest culprits to blame for declining ecosystems.

The worst driver of change is changes in land and sea use, followed by the direct exploitation of organisms, climate change, pollution and invasive species.

More than a third of the world's land surface and nearly 75% of freshwater resources are now devoted to crop or livestock production, which reduces the Earth's wild places and squeezes out native species.

The value of agricultural crop production has also increased by about 300% since 1970. Raw timber harvest has risen by 45%. Approximately 60 billion tons of resources are now extracted from the Earth every year - having nearly doubled since 1980.

The ocean has fared no better. In 2015, 33% of marine fish stocks were being harvested at unsustainable levels. Fertilizers entering coastal ecosystems have produced more than 400 ocean 'dead zones'.

The numbers can feel overwhelming to read – but they add up to a bleak picture.

Prof Andy Purvis, a biodiversity researcher at the [Natural History] Museum, was one of the scientists involved in the study.

He says, 'We should have gone to the doctor sooner. We're in a bad way. The society we'd like our children and grandchildren to live in is in real jeopardy.

'I cannot overstate it. If we leave it to later generations to clear up the mess, I don't think they will forgive us.

'One million animal and plant species threatened with extinction already - how much worse are we going to let it get?'

## Why we should care

All this loss in nature is now likely to translate into loss of human homes, economies and lives. As the natural environment is squeezed, over time it has less to offer Earth's human population, and without action, this problem will worsen.

The report found that up to 300 million people are already at increased risk of floods and hurricanes because of loss of coastal habitats and protection.

More than 800 million people face food insecurity in Asia and Africa, and about 40% of the global population lacks access to clean and safe drinking water.

There are currently more than 2,500 conflicts over fossil fuels, water, food and land occurring worldwide.

Loss of biodiversity is therefore shown to be not only an environmental issue, but also a developmental, economic, security, social and moral issue as well.

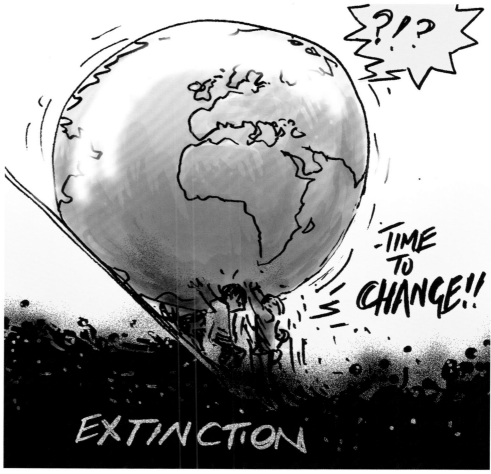

Audrey Azoulay, Director-General of UNESCO, says, 'This essential report reminds each of us of the obvious truth: the present generations have the responsibility to bequeath to future generations a planet that is not irreversibly damaged by human activity.

'Our local, indigenous and scientific knowledge are proving that we have solutions and so no more excuses: we must live on earth differently. UNESCO is committed to promoting respect of the living and of its diversity, ecological solidarity with other living species, and to establish new, equitable and global links of partnership and intragenerational solidarity, for the perpetuation of humankind.'

## What needs to be done?

The short answer is: a lot more than we are currently doing.

The report found that global goals for conserving and sustainably using nature cannot be met by current trajectories. Goals for 2030 and beyond may only be achieved through transformative changes across economics, politics and technology.

One of the most important ways we can create a sustainable future is by changing global financial and economic systems. According to the report, world leaders need to find a new way to build a global sustainable economy, steering away from the current limited paradigm of economic growth.

The researchers are calling for governments to include different value systems and diverse interests and worldviews when formulating policies and actions.

This includes listening to indigenous people and making biodiversity a priority.

Sir Robert Watson adds, 'We have already seen the first stirrings of actions and initiatives for transformative change, such as innovative policies by many countries, local authorities and businesses, but especially by young people worldwide.

'From the young global shapers behind the #VoiceforthePlanet movement, to school strikes for climate, there is a groundswell of understanding that urgent action is needed if we are to secure anything approaching a sustainable future.

'The IPBES Global Assessment Report offers the best available expert evidence to help inform these decisions, policies and actions - and provides the scientific basis for the biodiversity framework and new decadal targets for biodiversity, to be decided in late 2020 in China, under the auspices of the UN Convention on Biological Diversity.'

The full six-chapter report is expected to exceed 1,500 pages and will be published later this year.

*6 May 2019*

The above information is reprinted with kind permission from *NHM*
© The Trustees of The Natural History Museum, London 2021

www.nhm.ac.uk

# Biodiversity: we can map the biggest threats to endangered species in your local area

An article from *The Conversation*.

**THE CONVERSATION**

By Louise Mair, Research Associate in Biodiversity Conservation and Policy, Newcastle University & Philip McGowan, Professor of Conservation Science and Policy, Newcastle University

Since 1993, 15 species of bird and mammal are thought to have gone extinct, including China's Yangtze river dolphin and the Pernambuco pygmy owl from Brazil. But these recent examples are a tiny fraction of what scientists estimate could disappear in the lifetimes of people living today. One million species spanning the full diversity of life on Earth are at risk of extinction.

Trying to comprehend this scale of loss can make the problem seem insurmountable. Having a plan of action can help overcome that sense of powerlessness, and in new research, we've created one.

We developed a tool that can help governments, businesses and even members of the public discover how to halt wildlife extinctions. We worked with an international team of more than 80 conservationists to produce the Species Threat Abatement and Restoration (STAR) metric – a number that measures how much certain actions are likely to help reduce the extinction risk for local species.

## How it works

STAR uses data from the IUCN Red List of Threatened Species to give each species a score based on their conservation status. Species that are 'near threatened' according to the IUCN have a STAR score of 100, while species listed as 'vulnerable' have a score of 200. A higher score denotes a species facing a greater risk of extinction.

A critically endangered species, such as the Ka'apor capuchin in Brazil, has a score of 400. Breaking this down reveals which threats most contribute to the species' extinction risk, using data that quantifies their relative impacts. The greatest single threat to the Ka'apor capuchin is habitat loss due to expanding towns and cities. This contributes half of its extinction risk, and so accounts for 200 of the capuchin's points. Hunting and the selective logging of fruit trees, which this monkey forages from, make up the remaining 200.

STAR scores for different species living nearby can be added up to give the local area a total score. This represents a combination of how many species are present and how threatened those species are, and it can also be broken down to reveal which threats contribute the most to extinction risks for species in that area.

We applied STAR to all 5,359 amphibian, bird and mammal species on the IUCN Red List and found that halting the destruction of habitat for crop production would reduce their average extinction risk by 24%. Protecting habitats affected by the livestock industry would reduce their risk by a further 9% globally.

The expansion of agriculture plays a major role in biodiversity loss, but this doesn't mean that we should grow less food. Research has shown that combining more land-

efficient farming practices with efforts to protect and restore habitats nearby can feed the world's human population while conserving biodiversity. The STAR metric shows, at a 5km scale anywhere on Earth's land surface, where the negative effects of farming are likely to be particularly severe, revealing areas that urgently need action to halt habitat loss.

Threats vary between countries, as you might expect. Halting habitat loss from arable and livestock farms in Brazil would reduce the extinction risk of species nationally by 41%, whereas in South Africa, the figure is 17%. One of the major threats to wildlife here is invasive species. Controlling and eradicating non-native species could reduce extinction risk in South Africa by 15%.

### Tackling threats in biodiversity hotspots

Areas with very high STAR scores have lots of threatened species, and we might consider them particularly important for conservation. The country with the largest STAR score is Indonesia, where eliminating threats from farmland habitat loss, logging and hunting could reduce global species extinction risk by 7%. This is followed by Colombia (7%), Mexico (6%), Madagascar (6%) and Brazil (5%).

These five-highest scoring countries have much in common. In each, habitat loss due to crop production is the biggest threat and contributes at least a quarter of their national extinction risk. But in Brazil and Colombia, the next biggest threat is livestock farming, while in Indonesia, Mexico and Madagascar, it's logging and the timber industry.

There are schemes already in place in some regions to try to tackle these threats. In Indonesia, oil palm plantations can be certified sustainable if they meet environmental and labour rights standards. Expanding and effectively implementing these schemes could significantly reduce species extinction risk in these countries, potentially by as much as 30% in Indonesia.

### Local contributions to global conservation

While countries with high biodiversity have high STAR scores, wildlife conservation requires a global effort, and every country has an important contribution to make.

In the UK, there are over 30 birds and ten mammals threatened with extinction. Here in our home city of Newcastle in north-east England, the river Tyne hosts a particularly important breeding population of the kittiwake, while the bright-billed puffin breeds on the nearby Farne Islands.

Both of these seabirds are classified as 'vulnerable' by the IUCN. Overfishing of the puffin's prey, sandeels, contributes 25% to the species' extinction risk and a further 22% comes from climate change. This shows how important national and international policies are for strengthening local efforts to protect endangered species.

We can even use STAR to measure local and national contributions towards the UN Convention on Biological Diversity's 2030 goal of halting biodiversity loss, so that everyone can be part of the global plan for conservation.

*8 April 2021*

The above information is reprinted with kind permission from The Conversation
© 2010-2022, The Conversation Trust (UK) Limited

www.theconversation.com

# Wildlife trafficking driving 'severe declines' in traded species, finds study

Animals traded for pet industry, bushmeat, traditional medicine, ivory and lab use declined locally by up to 99.9%.

By Phoebe Weston

Wildlife populations decline by an average of 62% in areas where species are traded, pushing some closer to extinction, according to a new report.

The first analysis to quantify the impact of the legal and illegal wildlife trade looked at 133 land-based species and found the most endangered – which typically have smaller populations – are most at risk, with average declines of 81%. In some cases this resulted in local disappearances, with certain populations of spider monkeys and Baird's tapir declining by 99.9%, according to an international team of researchers led by Sheffield University.

Multiple local disappearances could lead to global extinctions, the research found. 'Our paper shows wildlife trade causes species to decline, which is a massive concern, because where species decline there is always a risk they could go extinct,' said lead researcher Oscar Morton, a PhD student at Sheffield University.

Some estimates suggest the illegal wildlife trade could be worth as much as $23 billion (£16.5 billion) a year, with more than 100 million plants and animals trafficked annually. The global impact of this trade on species in the wild was previously unknown. 'We reviewed thousands of published articles, in a huge comprehensive search of the available research. Then we analysed all this data from all these different species,' said Morton.

The team looked at the local and international wildlife trade, as well as legal and illegal trade. 'All trade leads to the same result – removing species from their habitat. Some illegal trade is sustainable but some legal trade is horrifically unsustainable. Here we wanted to assess the overall impacts,' said Morton.

The main drivers of wildlife trafficking are the pet industry, bushmeat (defined as wildlife traded for food consumption), traditional medicine, ivory and laboratory use. The study did not include subsistence-based bushmeat eaten by the communities that hunted it.

The researchers found only 31 studies that contained sufficiently rigorous data on population impacts, according to the paper, published in Nature Ecology & Evolution. These included 506 data samples containing population information on 99 species of mammal, 24 species of bird and 10 species of reptile.

Researchers compared areas where wildlife trade was active to unexploited control sites. They found that wildlife trade was driving population declines by 56%, even in protected areas. This research follows a study published in Science in 2019 which found 18% of the world's known land-based vertebrates are included in the wildlife trade, 50% more than previous estimates.

'All of these diverse forms of trade are suppressing wildlife abundance really dramatically,' said one of the paper's authors, David Edwards, professor of conservation science at Sheffield University, who described the findings as 'sobering'.

'The fact that we are seeing such severe declines over many different kinds of species and across different scales at which trade's occurring – I think that is a surprise. And I think it's something we all need to be really concerned about,' he said.

There was a lack of sufficiently rigorous data to include amphibians, invertebrates, cacti and orchids in the analysis, despite being significant parts of the global wildlife trade. There were also 'several alarming patterns' in the geographic coverage of suitable studies, with only four from Asia, one from North America and none in Europe, the researchers found. Most studies were focused on South America and parts of Africa.

'Lots of people who read this in the UK might not think this is anything to do with them. But it's about our wider relationship with wildlife, which we view as a replenishable commodity. If it's not proven to be sustainable, why do we assume it is?' said Morton.

National and international trade – which were found to be more significant drivers of decline than local trade – generally involve the extraction and trade of species of high commercial value, such as ivory from African elephants, horns from Javan rhinoceros and pangolin scales from across Asia and Africa.

Local wildlife trade involving the extraction or commercialisation of bushmeat supports an estimated 150 million households. Researchers say there is an urgent need for quantitative studies that support the potential for well-managed trade. 'Many hunters are likely already following sustainable practices and there needs to be widespread skill-sharing of these,' said Morton. The researchers said there should be better protective measures for threatened species and more research on the impacts of specific species at a local level. 'Improved management, tackling both unsustainable demand and trade reporting must be a conservation priority to prevent rampant trade-induced declines,' they wrote in the paper.

Dr Harry Marshall, a conservationist from Manchester Metropolitan University who was not involved in the research, said the methodology was robust and it was important to address the lack of research in the area.

'This research is important as it demonstrates quantitatively the impact trade is having on species on a global scale, which is potentially very large and concerning for certain species.'

Marshall said the impact of trade on population declines was predictable, but he was surprised that the study included legal trade. 'The impact of legal trade is often ignored and only recently being taken seriously, so it is good to see this covered,' he said.

*15 February 2021*

The above information is reprinted with kind permission from *The Guardian*.
© 2022 Guardian News and Media Limited

www.theguardian.com

# Five reasons people buy illegal wildlife products – and how to stop them

An article from *The Conversation*.

By Laura Thomas-Walters, Postdoctoral Research Fellow, University of Stirling, Amy Hinsley, Senior Research Fellow, University of Oxford & Diogo Veríssimo, Research Fellow in Conservation Marketing, University of Oxford

A British tourist on a tropical beach poses for a photo with a cute monkey-like animal. A Vietnamese man buys some rhino-horn powder and brags to his friends about its potency. An orchid collector admires their latest purchase, a stunning bright-pink flower, without worrying too much about where it came from.

These are very three different people, from different parts of the world. But what they all have in common is they are driving – sometimes unwittingly, often not – the illegal wildlife trade. This trade is one of the largest and most lucrative international crimes and poses a major danger to both wild populations and our own health. The complexity of the trade, which involves thousands of species sold for diverse purposes worldwide, can be a barrier to conservationists trying to design ways to address it. For example, reducing demand is vital, but this requires understanding why different consumers buy what they buy.

In our latest academic research, now published in the journal Conservation Biology, we have developed a classification of the common motivations behind the purchase of wildlife products. We found that, even in the midst of this complexity, wildlife consumers are driven by five overarching motivations:

### 1. Practical needs

Some wildlife consumers are motivated by everyday purposes. Get news that's free, independent and evidence-based

Can include:

Consuming wildlife for food or medicine; using wild-collected materials for building; using animals for labour; using wood or other plant material for fuel.

For example:

Dried seahorses are used to treat a variety of ailments in China and Taiwan, including sexual dysfunction, difficult childbirth and arthritis.

Potential solution:

Engage the traditional-medicine community to establish a voluntary code of conduct that focuses on sustainable trade, such as refusing to import small or pregnant seahorses.

### 2. Money

Some wildlife consumers are motivated by financial gain.

Can include:

Buying a wildlife product to profit from it, either immediately or as an investment for the future.

For example:

Vendors in Turkey and other tourist destinations use slow lorises as props in photo opportunities.

Potential solution:

Increase legal enforcement and sanction people caught peddling endangered wildlife. For instance, after the pop star Rihanna tweeted a selfie with a slow loris while on holiday in Thailand, local police raided a group of wildlife touts and made two arrests.

### *3. Social relationships*

Some wildlife consumers are motivated by social relationships.

Can include:

Trying to improve your social reputation; consuming due to pressure from family and friends; strengthening relationships with other people.

For example:

Rich businessmen in Vietnam buy expensive, illegal rhino horn to enhance their status among colleagues and clients.

Potential solution:

A campaign to persuade businessmen that success comes from within, that they do not need to rely on external products to impress others. (This is actually underway, see the ongoing Chi Campaign run by the NGOs Save The Rhino and TRAFFIC.)

### *4. The experience*

Some wildlife consumers are motivated by pleasure, novelty, or curiosity.

Can include:

Using wildlife products as part of a leisure activity; pleasing the senses with things that look, smell or feel nice.

For example:

Hobbyist orchid collectors in China value their beauty and care little if the plants have been illegally supplied from the wild. Consumer research shows that colourfulness is a major motivator.

Potential solution:

Breed legal, cultivated orchids to be more attractive to consumers (in this case, to be more colourful), thereby reducing the incentive to buy wild ones.

### *5. Spirituality*

Some wildlife consumers are motivated by spiritual needs, or the need to bring protection, luck or fortune in business and life.

Can include:

Using wildlife products to enhance spiritual wellbeing; religious practices; performing rituals or traditions.

For example:

Lansan tree resin is used to make incense for religious services in the Caribbean island of St Lucia. Overexploitation for resin means the tree is disappearing from the few Caribbean islands that make up its native range.

Potential solution:

Work with harvesters to develop sustainable harvesting techniques.

Both the products and the people involved in the wildlife trade are incredibly diverse, and consumers cannot always be pigeonholed into one neat category. In Indonesia, for example, songbirds are kept as pets. They are highly valued for their beauty and singing ability, and owners enter them into songbird competitions in the hope of winning both social status and cash prizes. In certain parts of Indonesia, the ownership of a songbird is also considered a rite of passage for young men.

People who buy songbirds may care about any or all of these reasons. However, using economic modelling techniques conservationists can discover which motivations are the most pressing and design effective interventions to halt this trade. By understanding what drives people to buy wild species, we can figure out how best to stop them.

*3 September 2020*

The above information is reprinted with kind permission from The Conversation
© 2010-2022, The Conversation Trust (UK) Limited

www.theconversation.com

# Species At Risk

**Chapter 2**

## Biodiversity crisis: quarter of UK's mammals at risk of extinction, first official Red List for country reveals

Wildcats, red squirrels and beavers among most threatened animals.

By Harry Cockburn

One quarter of the UK's native mammals are classified as being at 'imminent risk of extinction' and conservationists are calling for urgent action to save them, as the first official Red List for British mammals has been published.

The new list was put together by the Mammal Society and has received authorisation from the International Union for the Conservation of Nature (IUCN). It reveals 11 of the 47 mammals native to Britain are on the brink of extinction, while a further five species are classified as 'near threatened'.

Among those species listed as being at risk of extinction in Britain are the water vole, hedgehog, hazel dormouse, wildcat and the grey long-eared bat.

The European wolf is already extinct in the UK.

The reasons for the declines vary between species, the researchers said.

Many animals such as the wildcat, pine marten, and beaver have been subjected to extensive historical persecution.

For bats and the hazel dormouse, habitat loss is the main threat, while the water vole, red squirrel and Orkney vole suffer from the combined effects of habitat degradation and the introduction of non-native species.

Fiona Mathews, Mammal Society chair and Professor at the University of Sussex, led the report. She said: 'The new Red List provides a very clear basis for prioritising funding and conservation efforts for the future.

'Twenty species – those classed as threatened, near threatened, and data deficient – all need urgent attention. While we bemoan the demise of wildlife in other parts of the world, here in Britain we are managing to send even rodents towards extinction.'

She added: 'Things have to change rapidly if we want our children and grandchildren to enjoy the wildlife we take for granted.'

The UK is one of the world's most nature-depleted countries. The report says recovery for populations of declining species, as well as ensuring the survival of other creatures, will depend on returning large areas of land to nature.

Natural England chair Tony Juniper said: 'This is a wake-up call, but it is not too late to act. We are working with our partners to recover our threatened and widely loved mammals, including licensing the reintroduction of beavers into England, and supporting the recovery of dormice and the grey long-eared bat, but there is so much more to do.

He added: 'Central to the recovery of these and other creatures will be the protection and restoration of large areas of suitable habitat, including through the creation of a vibrant and wildlife-rich Nature Recovery Network, enabling populations of rare animals to increase and be reconnected with one another.'

The authors said reintroductions can offer hope for some species – reintroductions of beavers have been successful, with the animals readily breeding in the wild; and translocations of pine martens from Scotland, where over 98 per cent of the British population is found, have boosted populations in Wales.

Nevertheless, the animals will only cease to be classed as threatened once their populations are much larger and better connected.

But reintroductions are not a solution for most other threatened species because the causes of their declines have not been addressed.

'Instead, fundamental change is needed in the way we manage our landscapes and plan future developments, so that we provide the space and habitat needed for our wildlife to thrive,' the authors said.

Here is the list in full:

### Critically endangered:
- Wildcat
- Greater mouse-eared bat

### Endangered:
- Beaver
- Red squirrel
- Water vole
- Grey long-eared bat

### Vulnerable:
- Hedgehog
- Hazel dormouse
- Orkney vole
- Serotine bat
- Barbastelle bat

### Near threatened:
- Mountain hare
- Harvest mouse
- Lesser white-toothed shrew
- Leisler's bat
- Nathusius' pipistrelle

### Extinct:
- European wolf

The Red List for British Mammals has been produced by the Mammal Society on behalf of Natural England, Natural Resources Wales, Scottish Natural Heritage (NatureScot) and the Joint Nature Conservation Committee.

*30 July 2020*

The above information is reprinted with kind permission from *The Independent*.
© independent.co.uk 2022

www.independent.co.uk

# Only two northern white rhinos remain, and they're both female – here's how we could make more

An article from *The Conversation*.

By Ruth Appeltant, Postdoctoral Researcher in the Conservation of Endangered Species, University of Oxford & Suzannah Williams, Associate Professor in Ovarian Physiology, Lead for Ovarian Cryopreservation and Fertility Preservation Research, Lead of Rhino Fertility Project, University of Oxford

There were fewer than 100 southern white rhinoceroses (Ceratotherium simum simum) a century ago. Today, there are over 20,000. Sadly, this success story only stretches as far as the southern subspecies of the white rhino. With the death of the last male in 2018, the northern white rhinoceros (Ceratotherium simum cottoni) has passed the point where it can be saved naturally. With only two females remaining, the subspecies is now classed as functionally extinct.

This is a poignant, but not entirely hopeless, situation. New techniques, such as in vitro fertilisation (commonly known as IVF), enable us to bypass normal reproduction to produce new northern white rhino babies. Sperm samples from deceased males that are preserved in bio-banks solve one side of the equation, but there aren't frozen stores of northern white rhino eggs that we can rely on as easily.

We established the Rhino Fertility Project at the University of Oxford to help solve this problem. By using ovary tissue from deceased female rhinos to grow lots of eggs for fertilisation in a lab, we think we may have found a way to save the northern white rhinoceros – and potentially, other endangered species – from extinction.

## The first breakthroughs

A team led by Professor Thomas Hildebrandt from the Leibniz Institute for Zoo and Wildlife Research in Germany had a breakthrough in 2019 when they managed to collect eggs from the last remaining northern white rhinoceros females. After treating the females with hormones the immature eggs were collected, transferred to a lab where they were matured and then fertilised with frozen sperm.

To date, a handful of northern white rhinoceros embryos have been created this way. They're frozen and awaiting implantation in a surrogate female southern white rhinoceros. Transferring embryos into surrogates to produce baby animals is a process that's been well established for lots of species, including horses and cows, though it's still in the development phase for rhinos.

But the biggest constraint on this approach is that hormonal stimulation of female rhinos produces just a few eggs per cycle. Not all of these eggs will fertilise and not all will develop into an embryo. After transfer to a surrogate, only some will complete their development and become baby rhinos. As you might imagine, with only two remaining rhinos to gather these precious eggs from, this limits our ability to revive entire populations.

So what if we had the means to produce more eggs? While eggs collected from female rhinos are in short supply, generating eggs from ovarian tissue from deceased rhinos could fill the gap.

## Petri dish rhinos

As in humans, every female rhinoceros is born with thousands, if not millions, of immature ovarian follicles. At the centre of each of these sits an immature egg, also known as an oocyte. The follicle grows over months until it's ready to ovulate. At this point, these contain fluid and secrete hormones which influence the menstrual cycle. Ovaries contain lots of these immature follicles that are just waiting to be activated – in fact, far more follicles than are actually needed. As follicles grow and some are selected for ovulation, many are lost.

Rhinos don't undergo the menopause and so the ovaries of older animals still contain small follicles. Our goal is to grow these from the ovarian tissue of a deceased rhinoceros in the lab. We're experimenting with techniques that maximise the number of follicles we can grow in a culture dish.

By harnessing the full potential of rhino ovaries, we aim to grow as many eggs as possible. We're developing our method using ovarian tissue from different rhino species, including the southern white, Indian and black rhinoceros. Since all rhino species are either threatened or endangered, this technology could help more species than just the northern white, including the rare Javan and Sumatran rhinos. But as you can imagine, there aren't many rhinoceros ovaries available for laboratory research in the UK. Shipping tissue from threatened or endangered species in Africa to the UK in a timely manner is impossible, with many legal hurdles and mountains of paperwork. Collaboration between zoos, wildlife parks and research institutes is of the utmost importance here, to allow us to obtain this precious ovarian tissue.

The experience and knowledge we're gaining by developing this technique could even be useful in conservation efforts for other species. By freezing ovarian tissue and sperm from other endangered species and developing methods to cultivate follicles in labs, we could prevent further losses of some of Earth's most iconic wildlife and revive ecosystems rich in biodiversity.

*15 October 2020*

The above information is reprinted with kind permission from The Conversation
© 2010-2022, The Conversation Trust (UK) Limited

www.theconversation.com

# 'Really sad moment': bogong moth among 124 Australian additions to endangered species list

Ecologists say numbers declined by about 99.5% three years ago, likely due to drought, pesticides and light pollution.

By Lisa Cox

They were once so common, swarms of Australian bogong moths almost seemed to 'block out the moon' at certain times of the year.

Now, the bogong has been listed as endangered on the global red list of threatened species after crashes in its population in recent years.

The list, compiled by the International Union for the Conservation of Nature, was updated overnight with 124 new entries for Australian wildlife.

The addition of the bogong moth, famously seen in swarms at Parliament House in Canberra during its annual migration to the Australian alps, should be a wake-up call about declines in Australia's invertebrates, the scientist Marissa Parrott said.

Parrott is a reproductive biologist at Zoos Victoria and one of the researchers behind the moth tracker website that was launched two years ago to try to map the migration routes of bogong moths and learn more about changes in the species' populations.

Bogong moths were previously found in large numbers in parts of Queensland, New South Wales, the ACT, Victoria and South Australia.

'It is a really big step for the beautiful bogong moth to be listed as endangered,' she said.

'It is a really sad moment that a species that is so iconic to Australia, that people remember from their childhood as blocking out the moon, has now collapsed to the point of being listed as endangered.'

Other Australian species to be added to the red list include several other invertebrates, as well as plants and mammals.

Among them are Kangaroo Island species that suffered losses in the 2019-20 fires, such as the Kangaroo Island assassin spider, which has been listed as critically endangered, and the Kangaroo Island marauding katydid, listed as endangered.

The grey-headed flying fox, listed as vulnerable under Australian laws, has been given a vulnerable listing and the Arcadia velvet gecko, found in Queensland, has been listed as critically endangered.

Scientists have detected steady declines in numbers of bogong moths since the 1980s.

But in 2017 and 2018 that crashed to numbers so low the species was described as 'undetectable' in the alpine regions where it used to arrive in spring in numbers as high as 4.4 billion.

The moths are so important as a food source to animals such as the mountain pygmy possum that when they arrive in the mountains they are the second largest energy source after the sun.

Researchers working with Gunaikurnai Land and Waters Corporation have also highlighted the species' importance as a food source to traditional owners, with the discovery of microscopic remains of a bogong moth on a 2,000-year-old stone tool in a cave in Victoria's alps.

The ecologist Ken Green has been monitoring bogong moths for 40 years.

He and other researchers were consulted by the International Union for the Conservation of Nature as part of the assessment and asked if they could quantify the size of the declines.

'They said are we talking 60%? Or 40%? And we said no. Three years ago we had a decline of about 99.5%.'

Green recalls one set of surveys in the Canberra and Kosciuszko region in 2017 and viewing one cave that would typically house 17,000 moths a square metre. He said they could see just three moths inside.

Factors including pesticides and urban light pollution have been considered in relation to the decline in the species.

Green said Australia's drought through 2017, 2018 and 2019 was likely the largest contributor. He said last summer recorded a slight improvement in populations but numbers were, at best, 5% of what they used to be.

Jesse Wallace is a researcher at the Australian National University writing his PhD on the bogong moth. He has been studying their migration patterns for more than four years to try to learn how the species navigates to the same locations each year.

He remembers the first surveys he did in 2017 at breeding sites in NSW where in previous seasons hundreds of moths would have been caught every night.

He said the endangered listing was both sad and unsurprising.

'I caught about 50 in five weeks. That was the first indication that there was a problem,' he said.

Jess Abrahams, the Australian Conservation Foundation's nature campaigner, said the collapse of bogong moth numbers was affecting other species that rely on the moths for food.

'The bogong moth's population crash – and its cascading impact on other species – should concern every Australian, as we all depend on the interconnected web of nature, which gives us drinkable water, pollinated crops and clean air,' he said.

He added the State of the Environment report, published every five years by the federal government, was due to be released soon and was expected to reveal further declines in the health of plants, animals and ecosystems across Australia.

Parrott said sightings of bogong moths could still be recorded at the moth tracker website and people could help the species on its journey by keeping outdoor lighting to a minimum.

She said so far this year the website had recorded 150 sightings but only one swarm. 'It really should be a wake-up call that we need to help our invertebrates,' she said.

*9 December 2021*

---

The above information is reprinted with kind permission from *The Guardian*.
© 2022 Guardian News and Media Limited

www.theguardian.com

# New UK Red List for birds: more than one in four species in serious trouble

The latest assessment of the status of all the UK's 245 regularly occurring bird species – Birds of Conservation Concern 5 – shows that 70 species are now of 'highest conservation concern' and have been placed on the assessment's Red List. The newly revised Red List now includes familiar species, such as the Swift, House Martin and Greenfinch that have been added for the first time.

The report placed 70 species on the Red list, 103 on the Amber list and 72 on the Green list. Worryingly, the Red List now accounts for more than one-quarter (29%) of the UK species, more than ever before, and almost double the figure (36 species) noted in the first review in 1996. Most of the species were placed on the Red List because of their severe declines, having halved in numbers or range in the UK in recent decades. Others remain well below historical levels or are considered under threat of global extinction.

*Birds of Conservation Concern 5* is a report compiled by a coalition of the UK's leading bird conservation and monitoring organisations reviewing the status of all regularly occurring birds in the UK, Channel Islands and Isle of Man. Each species was assessed against a set of objective criteria and placed on either the Green, Amber or Red List – indicating an increasing level of conservation concern.

The report adds to a wealth of evidence that many of our bird populations are in trouble. Amongst the new additions to the Red List are the Swift, House Martin and Greenfinch.

Swift and House Martin have both moved from the Amber to the Red List owing to an alarming decrease in their population size (58% since 1995 and 57% since 1969

respectively). These join other well-known birds, such as the Cuckoo and Nightingale, already on the Red list, which migrate between the UK and sub-Saharan Africa each year. Work to address their declines must focus on both their breeding grounds here and throughout the rest of their migratory journey, which requires international cooperation and support.

The familiar garden bird, the Greenfinch has moved directly from the Green to the Red List after a population crash (62% since 1993) caused by a severe outbreak of the disease trichomonosis. This infection is spread through contaminated food and drinking water, or by birds feeding one another with regurgitated food during the breeding season. Garden owners can help slow transmission rates by temporarily stopping the provision of food if ill birds are seen and making sure that garden bird feeders are cleaned regularly.

Previous Birds of Conservation Concern reports have highlighted the plight of farmland, woodland and upland birds. This report adds more farmland and upland species to the Red List. Fifty-nine species of bird remain on the Red list from previous assessments; many of these, such as Starling, Curlew and Turtle Dove, are continuing to decline. As outlined in the 2019 State of Nature report, our bird populations face many pressures both here and abroad. These include changes in the way land is managed (particularly farmland which makes up 75% of the UK's land area), climate change, urbanisation, invasive non-native species and pollution.

The report also raises concerns over the status of wintering waterbird populations, with species such as Bewick's Swan joining the Red List. Pressures include illegal hunting abroad, the ingestion of lead ammunition, and the impacts of climate change. In addition, many of these wintering waterbird

populations have been affected by 'short-stopping', whereby they have shifted their wintering grounds north-eastwards in response to milder winter temperatures.

There is concern that the European wetlands they are now spending more of their time in may be drained or exploited in other ways and some are without protection altogether. Ensuring these areas are designated, protected and managed appropriately will become even more critical in safeguarding the ongoing survival of many of our migrating waterbirds.

The 2021 assessment does however contain some good news and demonstrates that targeted conservation action can make a real difference. The UKs largest bird of prey, the White-tailed Eagle, moves from the Red to the Amber List as a result of decades of conservation work including reintroductions and increased protection for this spectacular species. The population, however, remains low at just 123 pairs nationally. White-tailed eagles became extinct in the UK as a result of extensive habitat change combined, particularly in the 19th century, with persecution. Before their re-introduction, the birds last bred in England and Wales in the 1830s, in Ireland in 1898 and in Scotland in 1916.

The RSPB's CEO, Beccy Speight said 'This is more evidence that the UK's wildlife is in freefall and not enough is being done to reverse declines. With almost double the number of birds on the Red List since the first review in 1996, we are seeing once common species such as Swift and Greenfinch now becoming rare. As with our climate this really is the last chance saloon to halt and reverse the destruction of nature.

We often know what action we need to take to change the situation, but we need to do much more, rapidly and at scale. The coming decade is crucial to turning things around.'

The BTO's CEO, Prof Juliet Vickery, said; 'It is both sad and shocking to think that the House Martin, a bird that often, literally, makes its home under our roof, has become Red-Listed. As a long-distance migrant to Africa we know very little of its life outside of the UK, but possible causes include a lack of food, as a result of insect declines, and fewer nest sites from refurbishment of housing and the move to plastic soffits. Putting up artificial House Martin nest cups to provide safer nesting sites may not be the whole answer but it's a simple positive step many of us can take.'

The GWCT's Director of Research, Dr Andrew Hoodless, said: 'BoCC5 sadly adds more farmland and upland birds to the Red List. We need to better understand the effects of climate change on some species, as well as the impacts of changing habitats and food availability along migration routes and in wintering areas of sub-Saharan African migrants. For many Red-Listed species, however, improving breeding success in the UK is vital - we can and must make real and immediate improvements to this through better engagement with UK farmers, land managers and gamekeepers to encourage adoption of effective packages of conservation measures.'

JNCC's Director of Ecosystem Evidence & Advice, Steve Wilkinson, said: 'It is concerning to see three more long-distance migrants added to the UK Red List. We will continue to work with overseas partners to better understand the challenges faced by birds such as House Martin and Swift as they make their annual round trip between the UK and wintering grounds in Africa. Only through co-operation with countries along these flyways, can we hope to protect migrants like Bewick's Swan whose distribution is shifting in a rapidly changing climate.'

*2 December 2021*

The above information is reprinted with kind permission from *Rare Bird Alert*
© Rare Bird Alert 2022

www.rarebirdalert.co.uk

# Illegal ivory trade shrinks while pangolin trafficking booms, UN says

The demand for pangolin scales has increased tenfold, while the price of ivory has halved.

By Marcus Parekh

The illegal global ivory trade has decreased, but trafficking of pangolins is on the rise, a United Nations report into wildlife crime has revealed.

The study, compiled using four years of data, showed that revenue from ivory trafficking peaked between 2011 and 2013.

The United Nations Office on Drugs and Crime (UNODC) said that national bans on selling ivory, particularly the ban enforced in China in 2017, have caused the global trade to fall.

'The World Wildlife Crime Report 2020 has some good news and some bad news,' UNODC research chief Angela Me told Reuters.

'We see some shrinking in some markets, particularly the ivory and the rhino (horn) market, but we actually see huge increases in other markets, like the market of illicit trafficking of pangolins, in European eels but also in tiger parts and also in rosewood.'

Pangolins are a reclusive, nocturnal mammal covered in scales. Their scales are used in traditional Chinese medicine to promote blood circulation and reduce blood clots and there remains a large, legal market for this.

The causes in the shift in demand are opaque, due to the nature of the trade. The UNODC say the causes are multiple, but have cited the Chinese market as being the dominant driving force to the trade. Illegal poachers often catch the animals in Africa then smuggle them into Asia.

The annual income generated by the ivory trade between 2016 and 2018 is estimated to be $400m, down from over $1bn per year throughout the 1980s.

The UNODC say that a combination of market saturation and a shift in global attitude towards the trade have also contributed to a reduction in demand. Between 2014 and 2018, the price of illegal ivory in China halved.

By contrast, the seizure of illegally trafficked pangolin scales increased tenfold in the same period of time. The scales were primarily sourced from Africa and shipped to Asia. 185 tonnes of scales were seized during the four year window, which equates to approximately 370,000 animals being killed.

The UNODC say this shows that pangolins are now 'arguably the most heavily trafficked wild mammal in the world'.

*10 July 2020*

The above information is reprinted with kind permission from *The Telegraph*.
© Telegraph Media Group Limited 2020

www.telegraph.co.uk

# Chapter 3: Hope for the Future

## How hybrids could help save endangered species

An article from *The Conversation*.

By Lilith Zecherle, PhD Candidate in Conservation Biology, Liverpool John Moores University, Hazel Nichols, Senior Lecturer in Biosciences, Swansea University & Richard Brown, Professor of Animal Evolution, Liverpool John Moores University

What do you get when you cross two distinct lineages of an endangered species? For scientists hoping to revive an extinct population in Israel, the answer was a lucky accident – one that could upend longstanding ideas about how best to preserve biodiversity.

The Asiatic wild ass is a relative of the donkey that, as the name suggests, was never domesticated. This truly wild animal lives in the steppes and deserts of western and central Asia, from the Mediterranean to Mongolia. Because they vary slightly in size and colour (ranging from a pale sand colour to a dark ochre), Asiatic wild asses have been classified into five distinct subspecies: Mongolian, Indian, Iranian, Turkmen, and Syrian. The latter once roamed the Middle East, but overhunting drove it to extinction in the 1920s.

Hoping to return this species to the Negev desert, Israel started a reintroduction programme in 1968. The idea was not to resurrect the extinct Syrian wild ass, but to introduce an equivalent that was as similar as possible, which would fill the ecological role of the departed subspecies. And so, Israel imported 11 Asiatic wild asses from zoos in Iran and Europe.

Perhaps confusion regarding the species' taxonomy was to blame, but the 11 imported asses weren't all alike. In fact, they belonged to two different subspecies from Iran and Turkmenistan. And accidentally, a mixed breeding population was established in the HaiBar Yotvata reserve in Israel that created an entirely new hybrid ass.

Intentionally crossing two different subspecies risks creating offspring which are less healthy or viable. But, remarkably,

this accident seems to have paid off. In new research, we studied the genetics of the hybrid population and showed that mixing the subspecies might have actually helped make the reintroduction a success.

### What's wrong with hybrids?

Separate species and subspecies are genetically distinct. If they're too different, their hybrid offspring are likely to be much less healthy than the parents. Genetic incompatibilities between horses and donkeys mean that their hybrids – mules and hinnies – are infertile. Subspecies are more genetically similar than species and less likely to experience this problem, but there are other issues.

Different subspecies of the same species might have unique genetic adaptations that they've evolved in order to thrive within their particular environments. By crossing them, the next generation will carry fewer of these adaptations and might struggle as a result.

Imagine crossing a tiger from Siberia with one from Sumatra – the genetic make-up of the offspring may make them less comfortable than their parents in either snowy northern climes or tropical rainforests.

But mixing different populations can also help counteract the problems of small population sizes. Many endangered species persist in small and isolated populations and as a result, lose their genetic diversity through inbreeding and a random process called genetic drift. This spells danger for the whole species, as low genetic diversity means a lower potential for coping with sudden change, like a hotter climate or a new disease.

Think of it like playing scrabble, but your tiles contain just two letters. Your options are severely limited. A greater diversity of letters would give you much more flexibility to respond. Mixing subspecies seems to have given the reintroduced Israeli wild asses a more diverse genome.

### Return to the Negev

In Israel, the population of Asiatic wild asses grew rapidly from 11 founders in a mixed breeding group in 1968 to around 300 wild animals today. A fast-growing population is a sign of health. But starting with a small number of founder animals risks creating a reintroduced population with very low genetic diversity. Over time, this could cause it to shrink or even collapse.

We performed a DNA analysis to learn more about the genetic makeup of this population and its chances of persisting in the long-term. We found that all tested individuals of wild ass roaming the Israeli desert today are genetic hybrids and inherited half of their genes from each of the two subspecies.

This is more surprising than it sounds. If genetic incompatibilities between two subspecies usually make hybrids less healthy or fertile, we'd expect them to produce fewer foals than wild asses of pure subspecies ancestry. So, after several generations, we'd still expect to see some purebred asses, as they're presumably better suited to breeding healthy young than the hydrids. Yet we found no purebreds, suggesting there were no genetic barriers to interbreeding.

More importantly, the hybrid population became more genetically diverse over time. This may explain the population's rapid growth, and indicate that Israel's wild ass population is well-prepared for future challenges like disease and climate change.

### Diversity versus purity

The reintroduction in Israel is a rare success in the conservation of Asiatic wild asses. Other re-introductions have used only one subspecies and most of them have failed.

When 11 Turkmen wild asses were reintroduced to a reserve in Turkmenistan, the resulting population lost much of its genetic diversity over four generations. To date, only five out of 18 attempts to reintroduce wild asses have succeeded, including the one in Israel.

Despite the potential benefits, many conservationists oppose mixing different subspecies as it means sacrificing genetic purity, which some consider a precious component of a population's overall value. But our work suggests that focusing on genetic purity might not be the best strategy for saving an endangered species from extinction. Carefully assessing the potential risks is important, but hybrids can also form resilient populations with realistic chances of long-term survival.

*19 February 2021*

The above information is reprinted with kind permission from The Conversation.
© 2010-2022, The Conversation Trust (UK) Limited

www.theconversation.com

# How lockdown has helped the world's endangered species bounce back

With tourists absent, everything from coral reefs to rhinos has benefited.

BySarah Marshall; Andrew Purvis and Mark Eveleigh

Did 2020 have a silver lining? Sarah Marshall tells the story of wildlife recovery in numbers...

**1**

Eastern black rhino calf born in Grumeti Game Reserve, Tanzania, following the translocation of nine animals last year – a conservation triumph. The country's population has plummeted by 99 per cent since the 1970s to around 100 rhinos.

**3**

Wild red-and-green macaw chicks fledged in Argentina's Ibera National Park for the first time in more than 100 years. These vital seed dispersers are part of a bigger reintroduction programme masterminded by Rewilding Argentina Foundation, a partner of Tompkins Conservation, to save the native Paraná forest.

**7**

White-tailed sea eagles released on the Isle of Wight in August, at their last known breeding site used more than 240 years ago. Over time, the Roy Dennis Wildlife Foundation hopes that 60 of Britain's largest birds of prey will take flight.

**7**

Gorillas born in Bwindi Impenetrable Forest, causing the Uganda Wildlife Authority to announce a baby boom. There is no solid explanation for the unprecedented number, more than double that of the previous year.

**17**

Orphaned brown bear cubs rescued, rehabilitated and returned to the wild by the Orphan Bear Rescue Centre in Russia, supported by Born Free. Often abandoned as newborns, cubs are bottle-fed by biologists and taught how to survive in the wild.

**26**

Tasmanian devils reintroduced to the Australian mainland after an absence of several hundred – or possibly thousands – of years. Captive-bred by the Aussie Ark organisation, these meat-eating marsupials were released into a wildlife sanctuary in New South Wales with assistance from Avengers star Chris Hemsworth.

**48**

Ethiopian wolf pups born, boosting an ailing population of the world's rarest canid by a factor of 10 per cent. Just 500 individuals are spread largely between Ethiopia's Bale and Simien mountains. Disease, habitat loss and human persecution are the main culprits for the species' demise, but numbers like this could mark the beginning of a turnaround.

**58**

Sightings of critically endangered Antarctic blue whales off the coast of South Georgia by the British Antarctic Survey, heralding a return to their historic summer feeding ground. Only one was seen between 1998 and 2018, a legacy of the whaling industry boom of the 1900s.

**140**

Elephants born in Kenya's Amboseli Park, boosting further a population that has doubled in 30 years to reach almost 35,000. Good rains, improved anti-poaching efforts and a reduction in tourism are credited with this year's success.

**2,489**

Wild lions recorded in Kenya last year, according to a government report – an increase of 25 per cent since 2010. Across Africa, however, the king of the jungle is under threat, with a population of less than 20,000 – 10 per cent of what it was a century ago. The Kenya Wildlife Service has launched a 10-year action plan to conserve lions and spotted hyenas, which have been identified as the predators most at risk.

**4,000**

Grey seal pups crowding beaches at the National Trust's Blakeney National Nature Reserve in Norfolk. Just 25 were born in 2001. The success is down to low levels of disturbance and few natural predators. Forty per cent of the world's 300,000 grey seal population thrives in British and Irish waters.

*Good news: world wildlife in the spotlight*

Thailand's year of recovery

Bangkok, the world's most visited city, welcomed almost 40 million tourists in 2019. If a substantial number of those decided to forsake the crowded city and head to Thailand's unspoilt forests and sensitive reef habitats instead, the negative impact on wildlife could be huge. But the country believes it has stumbled upon a sustainable solution, largely as a result of the Covid-19 pandemic.

Animals in Thailand's national parks look set to enjoy the benefit of lessons learnt from almost a year of human absence – dubbed 'the anthropause'. The Department of National Parks has recommended that many of the country's parks and reserves close to the public for up to four months each year, mirroring the effect of the pandemic. Some parks are already forced to shut during the rainy season and, rather than have a rigid closed season, the idea is to stagger closures at various times.

'The absence of tourists leads to wildlife recovery,' said Thanya Netithammakun, director-general of the DNP, when the plan was unveiled. Already, Maya Bay in the Phi Phi Islands (which became a highlight of the backpacker route after it featured in the film The Beach) has been declared off-limits for the foreseeable future. So heavy was the footprint of mass tourism there, coral reefs might take 40 years to recover, says environment minister Varawut Silpa-archa.

In other heavily impacted areas, signs of change have already been witnessed as a result of the anthropause: hawksbill turtles have returned in record numbers to traditional nesting sites on the tourist beaches of Koh Samui; large herds of dugongs (sea cows) have materialised near the normally busy port of Trang; sea otters have been seen basking on Ranong beach, a popular haunt of backpackers; in Koh Chang National Park, starry puffer fish (measuring around 3ft long) returned after an absence of 13 years.

If the policy is implemented wisely, Thailand might have hit on a method for 'sharing the love' through less well-known parks while simultaneously diluting the pressure that comes with being an overexploited tourist gem. - Mark Eveleigh

### Undisturbed Galapagos... for now

The Galapagos Islands, always hailed as nature's paradise, could bring an even richer wildlife experience for visitors this year as a result of the pandemic. Pods of dolphins have revelled in empty harbours, brown pelicans have reclaimed nesting sites unused for decades, and bird populations have soared.

A census by the Galapagos National Park and the Charles Darwin Foundation revealed a record increase in flightless cormorants, plus a surge in the penguin population. Scientists largely credit La Niña – responsible for ocean cooling, and an abundance of food – but a drop in human disturbance no doubt helped.

So, would the remote Pacific archipelago be better off without us? Not at all. It appears prying eyes have a part to play in protecting vulnerable species.

Taking advantage of quiet waters, more Chinese fishing vessels than normal were spotted on the edge of the Galapagos last summer, raising fears of poaching activity. Residents of Santa Cruz island protested and activists led by Roque Sevilla – the former mayor of Quito, in mainland Ecuador – presented a marine protection strategy to the national government. Sevilla wants a 'multinational marine corridor' between Colombia, Panama and Costa Rica, protecting marine life from overfishing and other dangers.

On our imperfect planet, it seems animals need us as much as we need them. Tourism, conducted responsibly, keeps human predators at bay as well as providing income for conservation effort. - Sarah Marshall

### New lease of life for Australia's reef

Last month saw 'spectacular' coral spawning on Australia's Great Barrier Reef, raising hopes of recovery after years of damage by rising temperatures.

Most corals are hermaphrodites, producing sperm and eggs. When these are released, usually after a full moon, it results in what has been likened to a snowstorm of gametes underwater.

'We saw more corals spawning than in previous years,' said videographer Stuart Ireland, who filmed the event, while biologist Gareth Phillips said the release represented 'a genetic gold-mine of tough corals that have proved they can survive marine heatwaves.'

In 2016, 2017 and 2020, shallow corals were hit hard by 'bleaching' – the whitening that occurs when polyps are exposed to high water temperatures, often resulting in death. Once fertilised, the heat-resilient eggs released in December will have developed into larvae that are right

now drifting on currents and settling on the seabed to start new, more robust coral colonies.

Several dive operators in Queensland offer trips to see the annual spawning. - Andrew Purvis

### Pandas' baby boom

When the lights went out and everyone disappeared, giant pandas Ying Ying and Le Le stunned zookeepers at Hong Kong's Ocean Park by finally mating for the first time in 10 years.

Coincidence? Amy Cheung, from the park's public affairs department, likes to think so. She stresses that the park had closed many times before 'to minimise human disturbance' without any such aphrodisiac effect – but there is good reason to believe these clumsy bedmates were enjoying some rare privacy.

Although Ying Ying failed to conceive, Ai Bao and Le Bao at Everland amusement park in South Korea were more successful, giving birth to a 7oz pink nugget in July – a first for the country. There was more good news at the Smithsonian's National Zoo in Washington DC, where 22-year-old Mei Xiang became America's oldest giant panda to give birth. The conception was a little less natural (she was artificially inseminated) but it was a miracle that she managed to see the pregnancy to full term.

'Given her age, she had less than a 1 per cent chance of giving birth,' says Brandie Smith, the zoo's deputy director. A Giant Panda Cam has been set up to give viewers a closer look at the cub when travel returns to normal.

Female pandas come into heat just once a year, in March and April for a few days – which happened to be the start of lockdown last year. Who knows whether the absence of crowds had some bearing on the panda baby boom.

Whatever the reason, chief veterinarian Don Neiffer agrees that the results have created a feelgood factor. 'In the middle of a pandemic, this is a joyful moment we can all get excited about.' - Sarah Marshall

*30 January 2021*

The above information is reprinted with kind permission from *The Telegraph*.
© Telegraph Media Group Limited 2021

www.telegraph.co.uk

# Sharks will soon have their own 'superhighway' in the Pacific Ocean

Plans to create a 500,000 sq km marine reserve in the South Pacific could help whale sharks, turtles and other endangered species to recover, say conservationists

By Gavin Haines

A vast marine reserve is to be created in the Pacific Ocean, providing a protected 'superhighway' for hammerhead sharks, leatherback turtles and other endangered marine life.

The Eastern Tropical Pacific Marine Corridor (CMAR) is a joint conservation initiative between four Latin American countries: Ecuador, Colombia, Panama and Costa Rica. Unveiled at COP26 last week, it will see the nations link up and increase the size of their protected territorial waters, providing what they say will be a 500,000sq km sanctuary for species that have been hammered by overfishing.

The plan involves increasing the size of the Galapagos Marine Reserve – a crucible of life where Charles Darwin developed his theory of evolution – by 45 per cent. The expansion will encompass Cocos Ridge, an underwater mountain range between Galapagos and Costa Rica that is an important migration route for species such as whale sharks.

Galapagos Conservancy, a non-profit environmental organisation, described the move as 'a historic moment for Galapagos and a major victory for global marine conservation'.

Half of the expanded Galapagos reserve will prohibit all fishing, while the other half will permit longline fishing only. It is understood that fishing fleets will be prohibited from the rest of the CMAR. The challenge, as ever, will be enforcing the measures.

The Ecuadorian president Guillermo Lasso said that he didn't anticipate any resistance from his country's fishing industry. 'The proposal that I'm bringing here is a result of five months of negotiations with... the fishing industry and other sectors,' he said at a press conference. 'We made them understand the importance of this marine reserve.'

Lasso has not yet signed the decree for the establishment of the reserve, but is expected to do so imminently. A timeline for the creation of the expanded conservation zone has yet to be released.

The CMAR encompasses a region that attracts industrial fishing fleets, including shark-finning vessels, many from China. Galapagos Conservancy said it believed the protected zone, if properly policed, could allow sharks there to 'begin to rebound'.

'Although conservation work is never done, today we can celebrate a major victory for Galapagos and for our planet,' the organisation added.

*8 November 2021*

The above information is reprinted with kind permission from Positive.News
© Positive News

www.positive.news

# Beavers born in Essex for 'first time since Middle Ages'

Over 400 years after they were hunted to extinction, species has been reintroduced in county to tackle flooding.

By Harry Cockburn

Beavers reintroduced to tackle flooding in Essex have given birth to two kits - the first time the animals have been born in the county since the Middle Ages.

Spains Hall Estate in Finchingfield was the site where a pair of Eurasian beavers were released last year as part of a river management project and became the first pair to be brought to Essex in around 400 years.

The arrival of the babies means the reintroduction is off to a rapid start.

Beavers were hunted to extinction in the UK by the beginning of the 16th century due to demand for their meat, fur and scent glands.

Numerous reintroduction programmes are now underway to bring back the 'keystone species' which through their re-engineering of watercourses have dramatic positive ecological impacts on the landscapes they inhabit.

A five-year study into beavers which was completed earlier this year found Britain's wild beaver populations were reducing the impacts of floods, cleaning river water, and boosting populations of fish, amphibians and water voles.

Darren Tansley, river catchment coordinator at Essex Wildlife Trust, said: 'We always hoped that having beavers present would benefit the wildlife on site, but the changes we have mapped over the past 18 months have exceeded our expectations.

'DNA samples from the main beaver pond recorded everything from deer to tiny pygmy shrews and all this to create the perfect environment for their young kits, the first beavers born in Essex since the Middle Ages.

'We are thrilled by the addition of two more ecosystem engineers in the county.'

The adult beaver pair, Woody and Willow, have been building dams since their arrival as part of a partnership project with the Environment Agency and others.

Spains Hall Estate manager Archie Ruggles-Brise described news of the beaver babies as 'fantastic'.

'If they are anything like their parents, the two kits will become phenomenal dam builders, and we will be watching closely as they expand the wetland and provide even more protection against flood and drought, and provide homes for loads of other wildlife,' he said.

A public vote on Twitter is being held to help pick names for the kits.

*2 July 2020*

The above information is reprinted with kind permission from *The Independent*.
© independent.co.uk 2022

www.independent.co.uk

# Historic reintroduction reverses extinction of England's rarest frog

The northern pool frog, England's rarest amphibian, has been successfully reintroduced to Thompson Common, in Norfolk – reversing its disappearance from there in the 1990s. Thompson Common was the last site at which this species occurred prior to its extinction from England.

ARC staff released over 300 tadpoles into the reserve's pingos, post-glacial pools which offer the perfect habitat for the creatures. The tadpoles had been reared in captivity, away from predators, to increase the numbers of young frogs that survive through the tadpole stage.

This month's release saw the final phase in a process which began in 2015 to re-establish a population of northern pool frogs at their Norfolk home. This year's release is the fourth since 2015, bringing the total to more than 1000 tadpoles released. This year's release completes the reintroduction process. ARC's experience of species reintroductions is that three or four years of releasing animals helps to replicate a natural population structure with a varied range of ages.

The pool frog (*Pelophylax lessonae*) is assessed as critically endangered in England. The rare species went extinct in Britain in the mid-1990s, following the deterioration of its wetland habitat in the Fenland and Breckland areas of East Anglia. Its recovery has been made possible due to carefully planned, innovative conservation action by ARC and partners, including Natural England, Forestry England and Norfolk Wildlife Trust, and with support from Anglian Water, National Lottery Heritage Fund, Anglian Water Flourishing Environment Fund, Amphibian Ark the Keith Ewart Charitable Fund and the British Herpetological Society.

ARC Conservation Director Jim Foster said: 'It is not often that you can say that you brought an animal back from extinction, but that is exactly what we have achieved with our partners and with funding from the government's Green Recovery Challenge Fund.

'It is only because of the dedicated work here over the last few years that we have been able to bring it back. '

'The fantastic thing about Thompson Common is the sheer number and quality of the ponds here. Some of these glacial relic ponds, known as pingos, form ideal conditions for pool frogs.

'They are very open to the sun and they hold water all through the year, which is really important for the sun loving pool frog.'

Following the northern pool frog's demise, ARC and partners initially brought it back to a secret location in Norfolk in 2005, using frogs sourced from Sweden. The next step, to return them to Thompson Common, began in 2015, using tadpoles taken as spawn produced by the population which became established at the secret location.

The pool frog has not always been recognised as a native British species. Naturalists have long been aware of unusual frogs at Thompson Common and nearby. Within a few years, pool frogs jumped from being a non-native curiosity, to Britain's rarest amphibian and a conservation priority, to extinction. Although they were present within living memory, by the late 1990s, only a few years before recognition of native status, pool frogs had disappeared from Thompson Common, their final home. Research has shown that the native UK pool frogs are closely related to Scandinavian frogs, and have been present before reintroductions began and formed part of a distinct northern clade.

'The northern pool frog is very different to the common frog, which is the other native species of frog that we have in Britain. They look and behave differently – and they call loudly. Also, they lay fewer eggs, so they're much more vulnerable. A common frog can produce around 2,000 eggs, whereas a pool frog produces only about 500 or so.

Early indications at Thompson Common are good and we're confident that the frogs will form a self-sustaining population. The new population has begun moving around different areas of the common and has several hundred ponds to choose from' said Project Manager John Baker

Jon Preston, Conservation Manager for Norfolk Wildlife Trust, which manages the reserve, said: 'This is such an iconic species that fits in so well with the landscape we have got here.

'We are starting to see the sustainability of the population, they are breeding on site and we are seeing them spread further out from the release pools.

'It means the partnership, the head-starting and all the management work we do on site is all working.'

Working with a range of partners, until March 2022, ARC aims to continue to restore pool frog breeding ponds, train volunteers to survey for pool frogs, undertake specialist monitoring, bring pool frogs to a wider audience through commissioned videos, and plan future pool frog reintroductions.

This project is funded by the Government's Green Recovery Challenge Fund. The fund was developed by Defra and its Arm's-Length Bodies. It is being delivered by The National Lottery Heritage Fund in partnership with Natural England, the Environment Agency and Forestry Commission.

*2021*

The above information is reprinted with kind permission from The Amphibian and Reptile Conservation Trust
© 2022 The Amphibian and Reptile Conservation Trust

www.tarc-trust.org

# Dogs are saving endangered wildlife in Africa

Elephants, rhinos, pangolins and lions are under threat from poachers of endangered wildlife. Meet the dogs that sniff out contraband in Africa.

By Tira Shubart

Elephants, rhinos, pangolins and lions may not know it, but dogs are now their new best friends. They are called Canines for Conservation, and they help to protect endangered wildlife in six countries in Africa — six so far, but it's just a start.

The spectacular wildlife of Africa is under threat from poachers who hunt endangered animals to supply the illegal wildlife trade. International criminal networks sell ivory from elephants, horn from rhinos, teeth from lions and scales from pangolins to people who may not understand that an estimated 100 elephants are poached every day.

And at least two rhinos are poached daily. Lions, fewer in number than rhinos, have lost 85% of their habitat. And pangolins — the only mammal with scales — are the most trafficked animal in the world, with an estimated tens of thousands poached each year.

Some people mistakenly believe rhino horn and pangolin scales can be used as medicine. The truth is quite different. Horn and scales are made of keratin, the same substance as fingernails or the hooves of horses and cows. Lion teeth and claws are sought for jewellery in some Asian countries. Others want elephant ivory as a status symbol.

But now poachers in Africa, who hunt with military-style weapons and sometimes even helicopters, are up against dogs with a nose for detection.

## Dogs sniff out smuggled goods

Certain dog breeds, like Springer Spaniels and Belgian Shepherds, do battle with the illegal wildlife trade by stopping the smuggling of ivory, rhino horn, lion teeth and pangolin scales. The detector dogs can find smuggled goods in luggage and cargo at airports and ports across Africa, which disrupts the illegal business. The dogs are almost impossible to trick.

Spaniels and shepherds have found ivory and rhino horn even when it was wrapped in layers of metal foil or buried in

coffee and chili peppers. Dogs have sniffed out lion's teeth hidden in a thermos.

Once an illegal shipment is found, the finds are confiscated and the enforcement agencies start the detective work to track down the smugglers.

## How dogs fight crime against endangered wildlife

So how are the dogs and their human handlers trained to fight wildlife crime?

Will Powell, who heads Canines for Conservation, which is sponsored by the African Wildlife Foundation, is the man behind the scheme.

Powell has been training dogs his whole career, first teaching them to detect landmines in war zones. After helping to clear thousands of landmines on three continents, he focused on conservation in Africa. Powell loves his 'clever dogs,' and he chooses them carefully.

'We select dogs in places in Europe, where there is a culture of working dogs. We want dogs that have the canine equivalent of A levels, and then we take them further, sometimes to the PhD level,' Powell said, referring to the British exams before university and the highest university degree. The dogs Powell chooses are intelligent and tough, and can cope with the heat in Africa.

## Dogs have helped authorities arrest hundreds of smugglers in Africa.

Employees of wildlife authorities in Tanzania, Kenya, Uganda, Mozambique, Botswana and Cameroon are matched with dogs at Powell's base in Tanzania. Each human and canine team learn how to discover hidden wildlife contraband in eight to 10 weeks. The handlers learn to trust their dogs,

treat them well and give them time to play. Strong bonds of affection develop.

The dogs and their handlers have sniffed out contraband leading to hundreds of arrests of traffickers and the disruption of smuggling routes across Africa. There have been almost 400 seizures of illegal wildlife products at airports and sea ports since the programme started five years ago.

When the dogs make a successful discovery, they are rewarded with their favourite toy. And maybe on Earth Day, for their role in stopping endangered wildlife poaching, the detector dogs might be given an extra treat.

### THREE QUESTIONS TO CONSIDER:
1. There are eight species of pangolins, but each different species has the same tough, overlapping scales and tongues that can be half as long as their bodies. Why is this design so effective in helping them find their favourite food?
2. Elephants have almost the same age span as humans. But how long is an elephant gestation period?
3. What other aspects of nature in Africa do you think should be protected?

*22 April 2020*

*Tira Shubart is a freelance journalist and media trainer based in London. She has produced television news and trained journalists across four continents for international broadcasters, including BBC News, Canadian Broadcasting Corporation and Al Jazeera, over several decades. She is a Trustee and Co-Founder of The Rory Peck Trust for freelance journalists and an Ambassador for the Science Museum in London.*

The above information is reprinted with kind permission from NEWS DECODER.
© News Decoder 2022

www.news-decoder.com

# 200th osprey chick for pioneering Rutland Osprey Project

A ground-breaking project that reintroduced ospreys to England and helped bring them back to Wales has seen its 200th chick fledge this year.

Leicestershire & Rutland Wildlife Trust started the project 25 years ago because ospreys had become extinct in England and Wales. As a result of the project, ospreys have now spread across the two countries.

Ospreys are a huge fish-eating bird of prey with a wingspan of nearly 5 feet and can live for up to 20 years. The 200th chick, a female, fledged in July and was ringed with the number 360 to identify her.

Abi Mustard, Osprey Information Officer for the Leicestershire and Rutland Wildlife Trust, says:

'This year is an important and exciting year for the Rutland Osprey Project – we're thrilled to be celebrating our 25th anniversary and also welcoming the 200th chick. It's brilliant that we now have a self-sustaining population of ospreys in England.

'The success of the Rutland Osprey Project is not only due to the resilience of the birds themselves, but also to the hard work, support and dedication of everyone who has been involved – we have a wonderful team of volunteers, staff, local landowners and supporters who have helped facilitate these incredible achievements. We are all looking forward to seeing what the next 25 years brings.'

Leicestershire and Rutland Wildlife Trust launched the Rutland Osprey Project in 1996 in partnership with Anglian Water and the Roy Dennis Wildlife Foundation to reintroduce this magnificent bird of prey to the skies of England, where they had been extinct for over 150 years.

As well as establishing an osprey population in and around Rutland Water nature reserve, the project has helped the birds to breed in other parts of England and Wales.

Ospreys are now found breeding in Cumbria, Northumberland and North and West Wales, while Suffolk Wildlife Trust is working with the Rutland Osprey Project and Roy Dennis Foundation to bring breeding osprey back to East Anglia for the first time in over a century.

The Scottish Wildlife Trust has been helping to increase the number of ospreys in Scotland for over 50 years. Three of its reserves host breeding pairs. The Trust's Loch of the Lowes Wildlife Reserve has been at the forefront of the recovery of the species since 1969, when it became the fifth known breeding site in Britain. 85 chicks have fledged from the reserve in the past 52 years.

Ospreys were once widely distributed across the UK, but faced intensive persecution through shooting, egg collecting and habitat destruction, which eventually led to their extinction as a breeding species in England in 1847.

In the mid-1950s a population in Scotland began to slowly recover, however it was estimated that it would be approximately another 100 years before breeding ospreys would naturally recolonise central and southern England.

In a first, to help re-establish the birds to central England, the Rutland Osprey Project started translocating birds in 1996, carefully collecting 64 osprey chicks from Scottish nest sites and releasing them in Rutland between 1996 and 2001. A further 11 female birds were translocated in 2005. The first breeding pair of ospreys successfully raised a single chick at Rutland in 2001, and 25 years later, there are now approximately 26 adults including up to ten breeding pairs in the Rutland area.

2021 has brought another major milestone with the 200th chick, which hatched on a nest situated nearby on private land. The team hope she will return to Rutland to breed when she is mature.

Rob Stoneman, director of landscape recovery for The Wildlife Trusts, says:

'Seeing 200 chicks successfully hatch at the Rutland Osprey Project is a fantastic achievement. These beautiful birds belong in our skies, and it's thanks to the hard work of so many people over the last 25 years that we now have osprey across England and Wales.

'Success stories like this prove what's possible and help us to visualize how our countryside could look in the future – with wildlife in abundance, a rich tapestry of habitats, green corridors for species to move through landscapes, rivers and lakes free from pollution, and access to nature for all.'

This year's osprey chicks will likely remain in Rutland until early September, before they begin their remarkable 3000-mile migration journey south, to the west coast of Africa. The chicks will remain in their African wintering grounds for the first couple of years, so it won't be until at least 2023 before we see if the 200th chick returns.

Visitors to Rutland Water Nature Reserve can see ospreys by visiting the Lyndon Visitor Centre where two bird hides offer a exceptional view of the nest home to female osprey Maya and male 33(11), who have been breeding together at the reserve since 2015.

Plan your visit or watch the 24/7 nest webcam with a unique insight into the lives of a pair of breeding ospreys from the moment they return to their annual autumn migration.

*5 August 2021*

The above information is reprinted with kind permission from The Wildlife Trusts
© 2022 The Wildlife Trusts

www.wildlifetrusts.org

# The first endangered American animal has been cloned

By Vanessa Bates Ramirez

Last summer a horse named Kurt was born in Texas. Kurt wasn't just any horse—he was a clone made from DNA that had been frozen for 40 years and came from an endangered wild horse species from Central Asia.

Kurt was—and still is—pretty special. But now he's got some competition for the ti ftle of 'most amazing endangered animal cloned from frozen DNA.' The new contender is a black-footed ferret named Elizabeth Ann.

Elizabeth Ann was born in December in a conservation center in Colorado, the result of years' worth of careful research and planning. She's the first-ever endangered American animal to be cloned—and she may be her species' best hope for long-term survival.

## About the ferrets

Black-footed ferrets were endemic to Western US states like Wyoming and a Colorado, but started to die off in the mid-1900s when their main food source, prairie dogs, also died off due to disease and habitat loss. A small population of the ferrets was discovered in Wyoming in the 1980s, and skin biopsies from several of them were sent to San Diego's Frozen Zoo, where they went into a deep-freeze for over 30 years.

Scientists from conservation nonprofit Revive & Restore sequenced four of the ferrets' genomes using cells from the frozen biopsies, finding that genetic variation in the species had declined by an average of 55 percent since the 1980s. They chose the most genetically diverse male and female (Willa, the mother, had three times more genetic diversity than the average black-footed ferret) and sent their cells to a lab, which used them to create the embryo that became Elizabeth Ann.

## A unique embryo

How do you make an embryo from frozen skin cells, you ask? Scientists used an egg from a domestic ferret, the black-footed ferret's closest living relative. They removed the egg's nucleus and replaced it with the nucleus from one of Willa's cells. With a small electric zap used as an activation stimulus, the cells were able to divide as normal. This process is known as somatic cell nuclear transfer, and it's how Kurt the horse was born, as well as Dolly the sheep.

The embryo was implanted in a surrogate (non-black-footed) ferret, and the healthy clone of an endangered animal that lived almost 40 years ago was born in December. Revive & Restore calls the clone 'the most genetically valuable black-footed ferret alive.'

## Staying alive

As far as science goes, this is pretty cool stuff. Elizabeth Ann is the beginning of an effort to make her species more genetically diverse, thus increasing its chances of long-term survival; she'll soon be joined by other clones produced using the long-frozen cells, and they'll be mated and bred based on the strongest genetic outcome. They'll live at the National Black-footed Ferret Conservation Center near Fort Collins, Colorado, and could be reintroduced into the wild as soon as 2025.

'It will be a slow, methodical process,' said Samantha Wisely, a conservation geneticist at the University of Florida who was part of the project. 'We need to make absolutely sure that we're not endangering the genetic lineage of black-footed ferrets by introducing this individual.'

The ferret population will be closely monitored as new clones are born, but the outlook is positive. In a US Fish and Wildlife Service statement, Ryan Phelan, executive director of Revive & Restore, said, 'It was a commitment to seeing this species survive that has led to the successful birth of Elizabeth Ann. To see her now thriving ushers in a new era for her species and for conservation-dependent species everywhere. She is a win for biodiversity and for genetic rescue.'

*19 February 2021*

Vanessa is senior editor of SingularityHub. She's interested in biotechnology, genetics, renewable energy, the roles technology and science play in geopolitics and international development, and countless other topics.

The above information is reprinted with kind permission from SingularityHub. © 2022 Singularity Education Group.

**www.singularityhub.com**

# UK extinct species rediscovered in the Outer Hebrides

Buglife is delighted by the recent rediscovery of a species of caddisfly previously believed to be extinct across the UK. Once found in the fens of East Anglia, Limnephilus pati was presumed extinct in the UK in 2016 with no British records for over 100 years. Elsewhere, there are only 16 historical sites scattered across Denmark, France, Germany, Ireland, the Isle of Man and Poland.

In July last year, against all odds, a male was attracted to a light-trap being run by Robin Sutton on South Uist in the Outer Hebrides. Photos of the specimen were sent for identification and the exciting result was the rediscovery of Limnephilus pati.

South Uist is rich in habitats for caddisflies, with numerous small lochans, clear, low nutrient streams, and extensive machair habitats. Over the years Robin has attracted 23 species of caddisfly to his light trap but by far the most exciting find is Limnephilus pati.

Robin Sutton commented 'I've been running a light trap on the Outer Hebrides for over four years but I couldn't believe that the only location for a species thought to be extinct in Britain was in my back garden! It goes to show how much we still have to find out about these far-flung places.'

Craig Macadam, Buglife's Conservation Director, commented, 'It is really exciting that this species, thought to be extinct, has been rediscovered in the British Isles. We don't know a lot about its requirements, but the discovery of this new population means that we might be a step closer to working out what has driven the declines of this species elsewhere.'

5th March 2021

The above information is reprinted with kind permission from Buglife © 2022 Buglife

www.buglife.org.uk

# At least 28 extinctions prevented by conservation action

Conservation action has prevented the global extinction of at least 28 bird and mammal species since 1993, a study led by Newcastle University and BirdLife International has shown.

Newcastle University Press Office

## Extinction prevention

Publishing their findings in the journal Conservation Letters, an international team of scientists have estimated the number of bird and mammal species that would have disappeared forever without the efforts of conservationists in recent decades.

The species include Puerto Rican Amazon *Amazona vittata*, Przewalski's Horse *Equus ferus*, Alagoas Antwren *Myrmotherula snowi*, Iberian Lynx *Lynx pardinus*, and Black Stilt *Himantopus novaezelandiae*, among others. The researchers found that 21-32 bird and 7-16 mammal species extinctions have been prevented since 1993, with the ranges reflecting the uncertainty inherent in estimating what might have happened under hypothetical circumstances.

The study has highlighted the most frequent actions to prevent extinctions in these bird and mammal species. Twenty-one bird species benefited from invasive species control, 20 from conservation in zoos and collections, and 19 from site protection. Fourteen mammal species benefited from legislation, and nine from species reintroductions and conservation in zoos and collections.

The research team, involving experts from Sapienza University of Rome, Italy and the Zoological Society of London, among others, identified bird and mammal species that were listed as threatened on the International Union for the Conservation of Nature's Red List.

Led by Dr Rike Bolam and Professor Phil McGowan, from Newcastle University's School of Natural and Environmental Sciences, and Dr Stuart Butchart, Chief Scientist at BirdLife International, the team compiled information from 137 experts on the population size, trends, threats and actions implemented for the most threatened birds and mammals to estimate the likelihood that each species would have gone extinct without action. Their findings show that without conservation actions, extinction rates would have been around 3-4 times greater.

## Species for which extinction was prevented

One of the species the team evaluated was the Puerto Rican Amazon *Amazona vittata*, a small parrot species that lives on the island of Puerto Rico. The formerly abundant population was at its lowest in 1975, when only 13 individuals remained in the wild. Since 2006, efforts were made to reintroduce the species to the Rio Abajo State Park on Puerto Rico. In 2017, hurricanes wiped out the original population, only leaving the reintroduced population at Rio Abajo. Without the reintroduction efforts, the parrots would have gone extinct in the wild.

Other species remain only in captivity, but there is hope for such species to be reintroduced into the wild. The Przewalski's Horse *Equus ferus* went extinct in the wild in the 1960s. In the 1990s, reintroduction efforts started, and in 1996, the first individual was born in the wild. Now more than 760 Przewalski's horses are roaming the steppes of Mongolia once again. This gives hope that other species that are now only held in zoos, collections, or for plants in botanic gardens and seed banks, will be successfully returned to the wild in the future.

However, some species included in the study, such as the Vaquita *Phocoena sinus*, a species of porpoise, are still rapidly declining. While conservation may have successfully slowed declines, it may not be possible to prevent extinction in the near future without substantially greater resources, action, and political will.

Dr Rike Bolam, lead author of the study, said: 'It is encouraging that some of the species we studied have recovered very well. Our analyses therefore provide a strikingly positive message that conservation has substantially reduced extinction rates for birds and mammals. While extinctions have also occurred over the same

time period, our work shows that it is possible to prevent extinctions.'

Professor Phil McGowan, who leads an IUCN Species Survival Commission task force that provides scientific input into current negotiations on a new set of intergovernmental commitments on biodiversity and who is Professor of Conservation Science and Policy at Newcastle University, added: 'While this is a glimmer of hope – that if we take action we can prevent the irreversible loss of the last individuals of a species – we mustn't forget that in the same period, 15 bird and mammal species went extinct or are strongly suspected to have gone extinct.

'We usually hear bad stories about the biodiversity crisis and there is no doubt that we are facing an unprecedented loss in biodiversity through human activity. The loss of entire species can be stopped if there is sufficient will to do so. This is a call to action: showing the scale of the issue and what we can achieve if we act now to support conservation and prevent extinction.'

## New UN framework for tackling biodiversity loss

The findings are highly relevant to the UN Convention on Biological Diversity (CBD), which came into force in 1993. Over its lifetime, at least 28–48 bird and mammal species extinctions have been prevented. Through the Convention, governments adopted the 'Aichi Biodiversity Targets' in 2010, which committed countries to tackling the loss of nature. It is widely expected that the 5th Global Biodiversity Outlook – the CBD's official report due to be released on 15th September — will show that most targets have not been met.

The study provides a quantitative test of target 12, which aimed to prevent extinctions of known threatened species by 2020. The research found that 11–25 bird and mammal species extinctions were prevented over this timeframe, and that extinction rates would have been 3-4 times higher without action. Negotiations are now underway to develop a new framework to tackle biodiversity loss by 2030.

'These results show that despite the overall failure to meet the targets for conserving nature set through the UN a decade ago, significant success in preventing extinctions was achieved,' said Dr Stuart Butchart, Chief Scientist at BirdLife International, and instigator of the study. 'This should encourage governments to reaffirm their commitment to halt extinctions and recover populations of threatened species in the post-2020 Global Biodiversity Framework currently being negotiated. Such a commitment is both achievable and essential to sustain a healthy planet.'

*10 September 2020*

Reference
Bolam, F.C, Mair, L., Angelico, M., Brooks, T.M, Burgman, M., McGowan, P. J. K & Hermes, C. et al. (2020). How many bird and mammal extinctions has recent conservation action prevented? Conservation Letters, e12762. doi: https://doi.org/10.1111/conl.12762

The above information is reprinted with kind permission from Newcastle University.
© 2022 Newcastle University

www.ncl.ac.uk

# Key Facts

- About 98% of all the organisms that have ever existed on our planet are now extinct. (page 1)

- Land use change is continuing to destroy swathes of natural landscapes. Humans have already transformed over 70% of land surfaces and are using about three-quarters of freshwater resources. (page 2)

- Marine plastic pollution in particular has increased tenfold since 1980 – affecting 44% of seabirds. (page 6)

- About 80% of nitrogen used by humans – through food production, transport, energy and industrial and wastewater processes – is wasted and enters the environment as pollution. (page 6)

- Between 1970 and 2013, 56% of species in the UK declined. (page 8)

- Coral reefs are home to almost a quarter of all marine species. Roughly 33% of reef-forming corals are now threatened with extinction. (page 10)

- More than a third of the world's land surface and nearly 75% of freshwater resources are now devoted to crop or livestock production, which reduces the Earth's wild places and squeezes out native species. (page 10)

- Some estimates suggest the illegal wildlife trade could be worth as much as $23bn (£16.5bn) a year, with more than 100 million plants and animals trafficked annually. (page 14)

- The first official Red List for British Mammals, produced by the Mammal Society for Natural England, Natural Resources Wales, Scottish Natural Heritage (NatureScot) and the Joint Nature Conservation Committee, shows that 11 of the 47 mammals native to Britain are classified as being at imminent risk of extinction. (page 18)

- The latest assessment of the status of all the UK's 245 regularly-occurring bird species – Birds of Conservation Concern 5 – shows that 70 species are now of 'highest conservation concern' and have been placed on the assessment's Red List. The newly revised Red List now includes familiar species, such as the Swift, House Martin and Greenfinch that have been added for the first time. (page 22)

- Between 2014 and 2018, the price of illegal ivory in China halved. By contrast, the seizure of illegally trafficked pangolin scales increased tenfold in the same period of time. (page 24)

- Beavers were hunted to extinction in the UK by the beginning of the 16th century due to demand for their meat, fur and scent glands. (page 30)

- An estimated 100 elephants are poached every day and at least two rhinos are poached daily. Lions, fewer in number than rhinos, have lost 85% of their habitat. And pangolins — the only mammal with scales — are the most trafficked animal in the world, with an estimated tens of thousands poached each year. (page 32)

- The UN Convention on Biological Diversity (CBD) came into force in 1993. Over its lifetime, at least 28–48 bird and mammal species extinctions have been prevented. (page 39)

# Glossary

*Biodiversity*
Biodiversity is the name given to the amazing variety and variability of all life on Earth.

*Conservation*
Safeguarding biodiversity, attempting to protect endangered species and their habitats from destruction.

*Deforestation*
The clearance of large areas of forest to obtain wood or land for cattle grazing.

*Ecosystem*
A system maintained by the interaction between different biological organisms within their physical environment, each one of which is important for the ecosystem to continue to function efficiently.

*Endangered species*
A species of plant or animal at risk of becoming extinct.

*Environment*
The complex set of physical, geographic, biological, social, cultural and political conditions that surround an individual or organism and that ultimately determine its form and the nature of its survival.

*Evolution*
A gradual change in animals or plants over generations, during which they change their physical characteristics.

*Extinct*
If a species has become extinct, there are no surviving members of that species: it has died out completely.

*Groundwater extraction*
The process of extracting ground water from a source at a rate faster than it can be replenished.

*Habitat*
An area which supports certain conditions, allowing various species native to that area to live and thrive. When a species' natural habitat is mentioned, this refers to the area it would usually occupy in the wild.

*Hunting*
Hunting is the killing of animals for food or sport. Includes trophy hunting, where people kill animals and keep and display their body parts.

*Invasive species*
A species that is introduced into a new environment where it is not native and causes harm.

*IUCN Red List of Threatened Species*
The most commonly-used measure of how endangered a species has become is the IUCN Red List, which classifies endangered species as either Critically Endangered (CR) meaning that a species faces extremely high risk of extinction in the near future; Endangered (EN), meaning that a species faces a very high risk of extinction in the near future and Vulnerable (VU), meaning that a species is likely to become Endangered unless the circumstances threatening its survival and reproduction improve.

*Mass extinction*
A mass extinction event is when species disappear faster than they are replaced.

*Natural extinction*
If a species becomes extinct due to natural causes (as opposed to human causes such as poaching).

*Poaching*
Similar to hunting but the animals are killed illegally or without permission.

*Rewilding*
Reintroducing a species that previously inhabited a natural area.

*Species*
A specific type of living organism.

*Threatened species*
An animal which is at risk of becoming an endangered species unless the circumstances threatening its extinction change.

*Wildlife*
A collective term for wild animals and plants that grow and live independently of human beings.

*Wildlife trade*
The sale of wild animals, increasingly achieved through use of the internet to advertise and promote auctions.

# Activities

## Brainstorming

- Brainstorm what you know about endangered species:
  - Which species do you think are most at risk?
  - What are some of the causes of extinction?
  - What are the 5 biggest threats to UK wildlife today?
  - List as many species as you can think of that are already extinct.

## Research

- In small groups, choose a country and select an endangered species native to that country. Do some online research into the following:
  - the figures from the most recent count of that species
  - the major threats to that species
  - the efforts being done to preserve it

  Present your findings to the rest of the class.

- Do some research into the types of insects which are at risk of extinction. What is the causing the decline in their numbers? Write some notes on your findings and share with the rest of your class.

- In pairs, do some research into wildlife trafficking. Find out the following:
  - reasons why certain species are trafficked
  - reasons why people buy trafficked animals or their by-products
  - what efforts are being made to tackle the problem of the illegal wildlife trade

  Share your findings with the rest of your classmates.

- Research conservation organisations or charities in your local area and make a list of what species or habitats they are trying to preserve.

## Design

- Choose one of the articles in this book and create an illustration to highlight its key themes.

- In small groups design a poster to make people aware of the threat to bumblebees and what they can do to help protect them. Consider the impact their decline would have on plant life and the planet as a whole. Where would be the ideal places to display your poster?

- Select an animal from the critically endangered category on the WWF Species Directory. Create a social media campaign to raise awareness of the dangers your species faces and persuade people to contribute to a fund-raiser to help protect it.

## Oral

- In small groups discuss how you think climate change is threatening many species in the ocean and what you think needs to be done to change this.

- As a class, compile a list of extinct species and discuss which ones you would most want to bring back to life if you could, and why.

- Divide the class into two sides, for and against, and debate the motion 'Extinct species should not be revived'.

## Reading/writing

- Read the first article in this book and answer the following questions:
  - How many mass extinctions have there been in the past?
  - What were the names given to these mass extinctions and when did they occur?
  - What have been the causes of these mass extinctions?

- Write a one-paragraph definition of mass extinction and compare it with a classmates.

- Thinking about the research you have done and the articles you have read on this topic so far, write a list of the things you now know about endangered species that you didn't before.

- Write a list of things humans can do to alleviate species extinction.

# Index

## A
agricultural intensification 8
asteroids 2

## B
beavers 30
biodiversity 41
biodiversity loss 2–6, 10–11, 39
birds, at risk of extinction 13, 22–23, 34–35
bogong moth 21
bushmeat 14–15

## C
caddisflies 37
climate crisis 5, 8–9
cloning 36
conservation 7, 12–13, 15–23, 26, 29, 31, 33, 36–39, 41
coral reefs 10
COVID-19, impact on endangered species 27–28
Cretaceous mass extinction 2

## D
deforestation 2, 4, 41
Devonian mass extinction 1
dinosaurs 2
dolphins 12
drought 9

## E
Eastern Tropical Pacific Marine Corridor (CMAR) 29
ecosystem 41
ecosystem change 3–5, 10–11
elephants 32–33
extinction 1, 10
    natural 41
    preventing 38–39
    rediscovered species 37
    reversing 31
    species at risk 12–13, 18–21, 32
    see also mass extinction

## F
farming 8–9
ferrets 36
flooding 2, 9
fossil fuels 11
fossils 2
freshwater ecosystems 4–5, 10

## G
global warming 1
grasslands 3–4
groundwater extraction 4–5, 41

## H
habitats 41
heatwaves 5
hunting 4, 12, 13, 15, 22, 25, 41
hybrid species 25–26

## I
Intergovernmental Panel on Climate Change (IPCC) 5
Intergovernmental Science-Policy Platform on Biodiversity and Ecosystem Services (IPBES) 3, 10
International Union for the Conservation of Nature (IUCN) 4, 18
    Red List of Threatened Species 41
invasive species 2, 3, 7, 10, 13, 38, 41

ivory trade 24
    see also wildlife trade

## L
lions 32

## M
mammals, at risk of extinction 18–20, 36
marine life 28, 29
marine plastic pollution 6
mass extinctions 1–2, 41
metaldehyde 9

## N
natural extinction 41
natural resources, exploitation of 4–5
nitrogen 5–7

## O
Ogallala aquifer 4
Ordovician-Silurian mass extinction 1
osprey 34–35
over-abstraction 9

## P
pandas 28
pangolins 24, 32–33
Permian mass extinction 1
pesticides 9
pet industry 14
plastics 8
poaching 27, 28, 33, 41
pollution 2, 5–7
pool frog 31
prairies 3–4
pygmy owl 12

## R
Red List of Threatened Species 12, 21, 41
rhinoceros 20, 32
rivers 9
Rutland Osprey Project 34–35

## S
Scottish rainforest 6
Scottish Wildlife Trust 34–35
seabirds 13
seahorse 16
sixth mass extinction 2
Species Threat Abatement and Restoration (STAR) metric 12–13
synthetic biology 7

## T
tourism 27–28
trafficking of wildlife see wildlife trade
Triassic mass extinction 1–2

## W
white rhinoceros 20
wildfires 2
wildlife trade 14–17, 32–33, 41

# Acknowledgements

The publisher is grateful for permission to reproduce the material in this book. While every care has been taken to trace and acknowledge copyright, the publisher tenders its apology for any accidental infringement or where copyright has proved untraceable. The publisher would be pleased to come to a suitable arrangement in any such case with the rightful owner.

The material reproduced in **issues** books is provided as an educational resource only. The views, opinions and information contained within reprinted material in **issues** books do not necessarily represent those of Independence Educational Publishers and its employees.

### Images

Cover image courtesy of iStock. All other images courtesy Freepik, Pixabay & Unsplash except pages 34 & 35: David Tipling/2020 Vision & Abi Mustard & Andy Rouse/2020 Vision.

### Illustrations

Simon Kneebone: pages 18, 32 & 39. Angelo Madrid: pages 1, 29 & 39.

### Additional acknowledgements

With thanks to the Independence team: Shelley Baldry, and Jackie Staines. Contributing Editor: Tracy Biram

Danielle Lobban

Cambridge, January 2022

# Endangered Species

Editor: Danielle Lobban

Volume 402

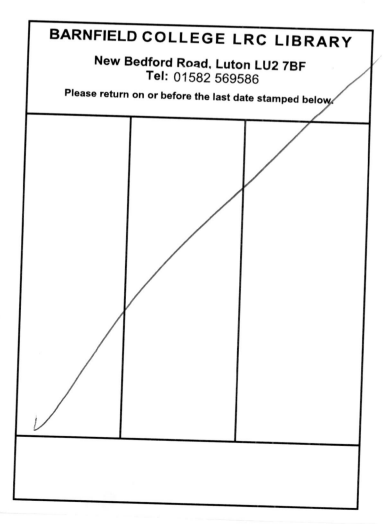

First published by Independence Educational Publishers

The Studio, High Green

Great Shelford

Cambridge CB22 5EG

England

© Independence 2022

## Copyright

This book is sold subject to the condition that it shall not, by way of trade or otherwise, be lent, resold, hired out or otherwise circulated in any form of binding or cover other than that in which it is published without the publisher's prior consent.

## Photocopy licence

The material in this book is protected by copyright. However, the purchaser is free to make multiple copies of particular articles for instructional purposes for immediate use within the purchasing institution. Making copies of the entire book is not permitted.

ISBN-13: 978 1 86168 861 3

## Printed in Great Britain

Zenith Print Group